AS
LEVEL

GEOGRAPHY
FOR CCEA AS LEVEL

2nd EDITION

D1332547

COLOURPOINT EDUCATIONAL

Rewarding Learning

**Martin Thom and
Eileen Armstrong**

COLOURPOINT
EDUCATIONAL

Colourpoint Educational
An imprint of Colourpoint Creative Ltd
Colourpoint House
Jubilee Business Park
21 Jubilee Road
Newtownards
County Down
Northern Ireland
BT23 4YH

Tel: 028 9182 0505
Fax: 028 9182 1900
E-mail: sales@colourpoint.co.uk
Web site: www.colourpointeducational.com

The Authors

Martin Thom is a teacher of geography, an Examiner
for an awarding body and a writer of articles and
textbooks. He would like to thank his colleagues and
pupils at Sullivan Upper, both past and present, for
helping him to communicate how geography uniquely
links people to both their everyday world and the
extraordinary.

Eileen Armstrong has a BA (Hons) from QUB and has
taught in a number of leading Grammar schools in
the greater Belfast area. She has worked as a Senior
Examiner for an awarding body for over 20 years. She is
co-author of a number of endorsed A level Geography
text books published by Colourpoint.

The author acknowledges her husband's contribution
to the research, data collection and photographic
material used in this book. She also acknowledges the
help and support of the editorial and design staff in
Colourpoint, Newtownards.

Rewarding Learning

Contents

Unit AS 1:
Physical Geography

> "Everything has to do with geography."
> *Judy Martz*

Students should be able to:

(i)   explain how the drainage basin operates as an open system with inputs, outputs, stores and transfers of energy and matter

(ii)   understand storm and annual hydrographs (regimes) and explain the factors that influence them – relief, basin size and shape, soil, geology, land use, drainage density and precipitation

(iii)   understand river processes – erosion (abrasion/corrasion, attrition, hydraulic action, solution/corrosion), transportation (suspension, solution, saltation, traction), and deposition (Hjulström curves)

(vi)   explain the formation of river landforms – waterfalls, rapids, meanders, pools and riffles, oxbow lakes, levées, floodplains and deltas (arcuate and bird's foot)

Few subjects touch the world in which we live quite like Geography. Listen to any news bulletin and topic after topic relates to some aspect of this multidimensional subject. So all-embracing is Geography that it defies easy definition, leaving some to state: "Geography is what Geographers do". This textbook, like the AS specification on which it is based, attempts to illustrate the unity in diversity of the subject. While its structure reflects three aspects of both physical and human geography, the overall theme is of interaction and understanding of the world through skills.

Seeing the world geographically – a systems approach

Making sense of the world has been a universal goal across time and space. The need to create order from the night sky initially led to the identification of patterns and constellations representing earth-bound items: lions and ploughs or mythical wonders, gods and giants.

One approach to unifying the study of geography is known as **systems theory**. In essence this is simply a pattern or framework through which aspects of the world can be viewed.

As illustrated (Figure 1) a system exists where a number of objects are linked by interactions between them. The objects are known as **stores** and the interactions as **transfers**. To identify a system the starting point is to recognise its boundaries. A computer has clear boundaries and even the most advanced computer needs input such as power and data to operate. A system is said to be **isolated** if nothing transfers in or out across its boundaries or **closed** if only energy can enter or leave it. Most natural systems are **open,** as both materials and energy enter (**input**) and leave (**output**) them. The whole universe and its numerous parts can be regarded as systems. This can help us simplify the world and aid our understanding of how it operates.

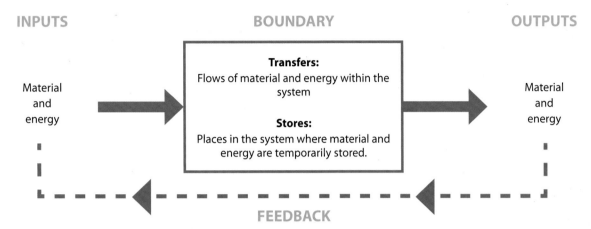

Figure 1: The elements of an open system

The drainage basin as an open system

Fluvial environments are those that result from the action of river processes.

The area of land that gathers water for a river is known as a **drainage basin**. The land boundary of a drainage basin is its **watershed** – an area of higher ground separating drainage down one slope from another. When **precipitation** (rain, hail or snow) falls from the atmosphere onto the land it must either return to the air (evaporate), sink into the ground or flow over the surface. The vegetation, ground surface, soil, underlying rocks and river channels are all **stores** in the basin. Infiltration, overland flow, through flow and groundwater flow are some processes or **transfers** within the basin system. Transfers of energy and materials across the system's boundaries may be either **inputs** – including precipitation and potential energy – or **outputs** – such as evapotranspiration and runoff (Figure 2). Most natural river networks have many **channels** that join together as **tributaries** (contributing water) to form larger rivers and drainage basins.

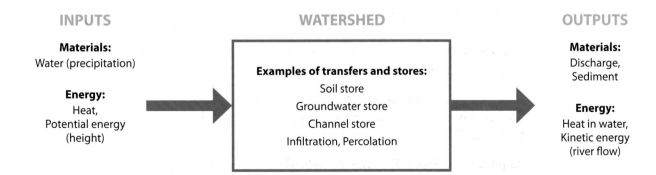

Figure 2: The drainage basin as an open system

While water is the most obvious component that moves in, through and out of a drainage basin open system, there are others including sediment and energy (heat and potential energy). Figure 3 outlines the transfers of water across the boundaries of a drainage basin and between its stores.

Figure 3: Movement of water into, through and out of the drainage basin

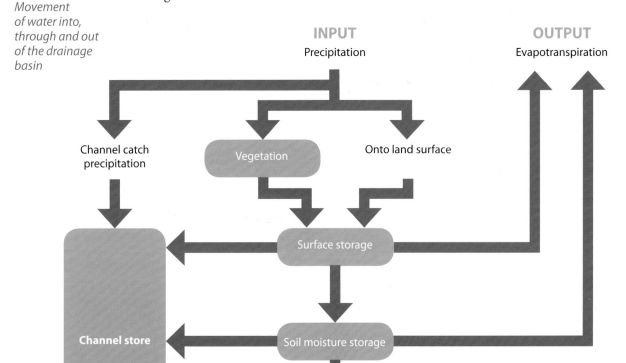

Processes in a drainage basin

1. Surface flows and outputs

Precipitation may enter a drainage basin directly into the river itself (channel catch), onto the land surface, or caught on the vegetation (interception). Once it reaches the surface, water may be lost to the system by evaporation or it may move across the land (surface runoff or overland flow) or into the soil (infiltration). Once in the soil, water may be taken up by plant root systems and ultimately lost through leaves back to the air (transpiration). The output from a drainage basin into the atmosphere is termed **evapotranspiration**. Water that flows over the land will gather in the river channels at the base of slopes.

2. Infiltration and subsurface flows

One of the key factors in how a river basin responds to rainfall is the movement of water into the soil and through the soil and groundwater stores. Soil is mostly made of inorganic particles (rock fragments) of various sizes, from fine *clays* through medium *silt-sized* to large *sand-sized* material. The exact mixture of these particles in a soil is described as the **soil texture** (Figure 4).

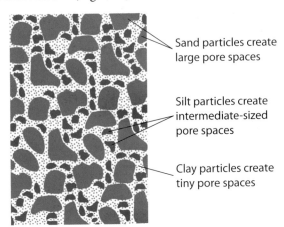

Sand particles create large pore spaces

Silt particles create intermediate-sized pore spaces

Clay particles create tiny pore spaces

Figure 4: Illustration of pore space in a soil with mixed texture

Soil particles are separated from each other by small spaces or pores. It is through these that water can:

- infiltrate from the surface into the soil.
- percolate from the soil down into the underlying rock.
- through-flow downslope within the soil.

Different soils have different rates of infiltration, for example, sandy soils have large pores and water moves in quickly. This characteristic is the soil's infiltration rate and this is measured in mm per hour.

Soils that allow water to enter rapidly are described as **permeable** whereas **impermeable** soils, often clay-rich, allow water to enter only slowly. A key factor to note about infiltration is that the rate changes during the course of a rainfall episode (storm). As Figure 5 shows, no matter what soil type is found, the infiltration capacity is relatively high when rain starts. This is because the soil is dry and has pore space available, but the infiltration capacity falls to a lower, stable level as the pores fill up over time.

Figure 5: Typical infiltration rates for different soil types

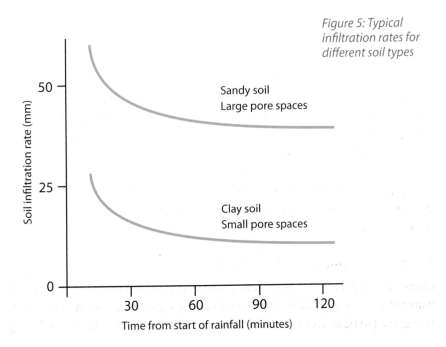

Sandy soil
Large pore spaces

Clay soil
Small pore spaces

Soil infiltration rate (mm)

Time from start of rainfall (minutes)

The movement of water from the soil into the underlying rock is called **percolation**. Most rocks can store water either within the rock itself (porous), or within the cracks and joints they contain. This stored water is referred to as **groundwater** and it slowly makes its way into river channels as **groundwater flow**. The height of the groundwater level in rock is known as the **water table**. This underground source of water can maintain river flow even after weeks or months without rainfall. Remember river systems also transfer sediment as well as water.

Measuring and recording river flow

Discharge is the volume of water passing any one point in a given time. It is calculated by multiplying the channel's cross-sectional area (metres squared) by the river's velocity (metres per second). The units of discharge are therefore cubic metres per second ($m^2 \times m/s = m^3/sec$) or **cumecs**.

Figure 6: How to find the discharge of a river

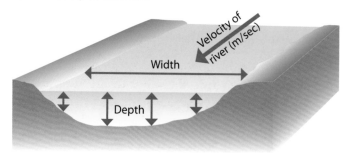

The pattern of changing discharge at any point along a river can be plotted on a time/discharge graph called a **hydrograph**. An **annual hydrograph** would show the variation in river discharge over a year whereas a **storm** or **flood hydrograph** records the impact of one period of precipitation on water flow.

Annual hydrograph

A river system's annual hydrograph shows its variation in discharge over a year. Most usefully these can be averaged, over 30 years or more, to give the mean flow pattern (Figure 7). Just as the weather describes the day-to-day change of the atmosphere and climate describes the normal pattern of seasonal weather, the Annual Hydrograph, sometimes called a river's regime, illustrates the system's normal yearly flow pattern.

Figure 7: The annual hydrograph or regime of the River Severn

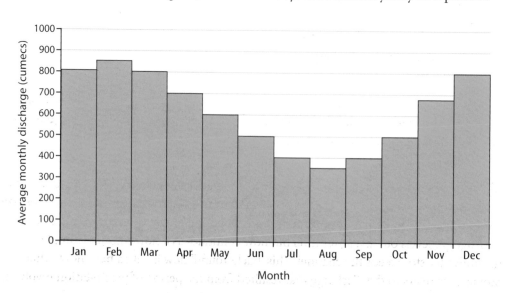

Events such as spring snow melt in mountain regions or the change between wet and dry seasons in the Asian monsoon lands will be reflected in the shape of annual hydrographs. Across much of Europe the river year is taken from the 1 October to the following 30 September. This is because rivers are normally at their lowest flow level in Europe around those dates.

Why would this be likely to be true in the UK or Ireland?
Clue: It is not about the amount of precipitation it has more to do with other ways that water leaves (outputs) drainage basins.

Annual hydrographs are useful for planners when they are examining new developments or changing land use across a basin. They are also important for water engineers working to ensure an adequate year round water supply and prevent water shortages or floods.

Storm hydrograph

Figure 8 shows a model storm hydrograph, including the terms used to describe the line of changing discharge, and the area beneath the line that represents the total volume of channel flow. The superimposed bar graph shows the amount and pattern of precipitation (storm) that causes the subsequent changes in flow. Two sets of transfers

Figure 8: Model hydrograph of storm flow

within drainage basins impact the shape and nature of a hydrograph: firstly, the **fast (surface) flow** processes of channel catch and overland flow; and secondly the **slow (sub-surface) flows** of through flow (soil) and groundwater (rock) flow.

A typical storm hydrograph follows a sequence. After rainfall commences, some falls directly into the channel and discharge rises, but much of the rain falls on the basin and may infiltrate into the soil. Eventually the soil becomes saturated, or if the rainfall is intense and exceeds the soil's infiltration capacity, water will flow over the surface as overland flow and quickly reach the channel. This accounts for the initial increase in discharge shown as the **rising limb**.

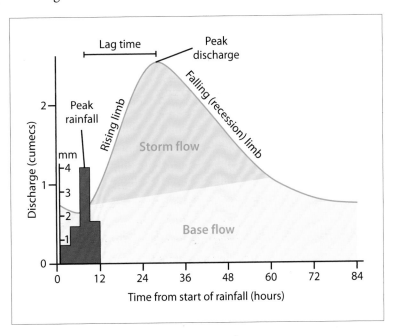

Even in small drainage basins it takes time for water to make its way to the channel and this delay between the **peak rainfall** (bar graph) and the **peak discharge** (hydrograph) is called the **lag time**, and may be measured in minutes or days depending on size of the drainage basin.

The peak discharge is a critical point because if it higher than the carrying capacity of the river channel, flooding will occur. Following the peak discharge is the **falling** or **recession limb**, where the remaining water from the storm period moves from the basin. The hydrograph levels out at a lower level but flow continues as a slow transfer from the groundwater store feeds the channel. This 'background' flow is called **base flow**, while **storm flow** refers to the discharge that resulted from the period of precipitation involved.

In general terms the shape of hydrographs varies between two extremes:

- A wide, shallow curve with gentle rising and recession limbs, and a low peak discharge following a long time lag. This is known as a **flat response**.
- A steep-sided, rapidly changing graph with a high peak discharge and a short lag time. This form is known as a **flashy response** and often brings the threat of river flooding.

Figure 9:
Hydrograph shapes

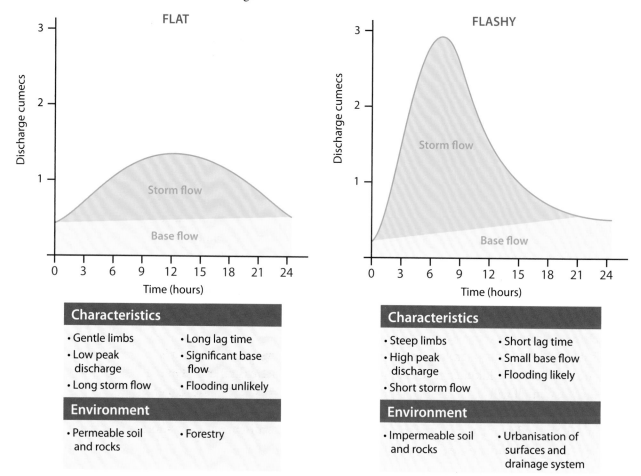

Factors affecting basin discharge and the storm hydrograph

The way that a basin responds to rainfall, as shown by its flood hydrograph, depends on many different factors. These can be classified in two groups: 1. the nature of the storm or 2. the nature of the drainage basin.

1. Nature of the storm

A period of precipitation has many variables, all of which can influence how drainage basins deal with the input. How long the storm lasts, how heavy the storm is, the type of precipitation (rain, hail, or snow) or even the direction and speed of the rain storm across the basin will all impact on the hydrograph shape. It is fair to say that no two storms are identical.

Try to imagine how each of these factors would change the appearance of the storm hydrograph. What would be the impact of a long period of light rain compared to an intense short but heavy storm? Would the impact of identical storms be different if one took place in February and the other in August? In your answers use as many of the hydrograph terms as possible.

2. Nature of the basin

a) Permanent features

Basin size and shape

In a small basin of a few square kilometres, rain will be gathered and transferred quickly producing a short period of storm flow and a short lag time. In larger basins, such as the Nile or Congo, the precipitation may take days or weeks to flow through the basin. The most efficient shape for a basin would be circular with a central outflow; this is unrealistic but long narrow basins will produce longer, more even-shaped hydrographs compared to the flashy, steep-sided response of roughly circular basins.

Relief

Naturally steep basin slopes will shed water more rapidly than gentle slopes. For this reason, drainage basins in mountain areas will produce more dramatic flashy hydrographs in comparison to those found in low-lying basins where gentle slopes slow down transfers.

Underlying soil and geology

The ability of soil and the underlying rock (geology) to take in and store water will have a fundamental impact on hydrographs. The limestone and chalk rocks of Southern England and Central Ireland are highly permeable, reducing river channel flow levels. In regions of impermeable soil (clay) or rock (granite) the drainage basin's storm hydrograph will have a flashy response to rainfall as fast surface flows are dominant.

Drainage density

This measures the amount of river channel within a drainage basin area and is often expressed as kilometres of channel per square kilometre (km/km^2). On the surface rainfall will drain rapidly if it can reach a channel quickly, so the higher the drainage density, the more efficiently the basin will be drained. This will result in a flashy hydrograph with a steep rising limb, a high peak discharge and a short lag time. (For summary, see Figure 10.)

b) Temporary features

Previous (Antecedent) conditions

This is the state of the basin just before a storm event. For example, if there has been recent rainfall in the area that may have saturated the soil or alternatively if a period of drought left the soil and groundwater stores depleted and able to take in and store a large proportion of the precipitation.

Vegetation cover

A basin covered in vegetation will have a high evapotranspiration output, so reducing the total amount of precipitation that is transferred by the river itself. The type of vegetation, grass or trees, and seasonal variation will have a strong impact on the shape of hydrographs. In summer, deciduous trees intercept 60% more precipitation than in winter and the rainfall that infiltrates the soil is more likely to be taken up by tree roots for growth than to reach the river channel directly. A single mature oak tree uses up to 230 litres of water each day in the summer growing season.

c) Land use change – the human factor

Vegetation change

Human activity alters the nature of the land surface, not only the obvious development of urban landscapes, but even the vegetation cover of the remote rural landscape. In the mountains of central Wales two rivers, the Severn and the Wye, have their sources in adjacent basins on the slopes of Plynlimon mountain. In many respects the two basins are very similar in size, shape and underlying geology. They are also subject to very similar rainfall conditions. However, since the 1950s one has been extensively used to grow coniferous forest on a commercial basis. The other basin has a vegetation cover of upland heath and moor used for hill sheep farming. The outcome of this difference in land use is seen in the respective hydrographs for the two basins (Figure 11). In addition, different types of farming such as growing grassland pasture or planting arable crops will alter the amount and speed of water flowing through a river basin.

How might a field of grass give a different response to rainfall than a similar field ploughed ready for wheat to be sown?

Exercise

1. Select **three factors**, one from each column in Figure 10. Discuss, in detail, how each factor can influence water transfer in a drainage basin. Using appropriate terminology, describe and explain how each factor could impact the shape of a hydrograph.

Figure 10: Some factors influencing basin hydrographs

Nature of the storm	Nature of the basin 1 *Permanent features*	Nature of the basin 2 *Temporary change*
Total precipitation Type of precipitation (rain/snow) Intensity of precipitation Duration of storm Path of storm across basin Frequency of storm Speed of storm movement	Size of basin Shape of basin Aspect of basin Slopes in basin Drainage density (length of channel per unit area) Channel size Nature of soil Nature of rock	Vegetation (seasons) Land use (farming/urban) Previous weather Soil moisture Groundwater storage

2. With reference to Figure 11, compare the response of the Severn and Wye river basins to the same rainstorm. Using figures from the graphs and the correct hydrograph terms (Figure 8), explain the difference between the two hydrographs.

Figure 11: Hydrograph and catchment map of Upper Severn and Wye river basins

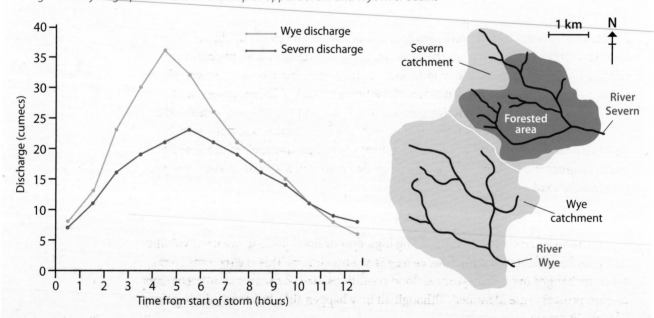

Urbanisation

While the planting of trees – afforestation – modifies basin response to a less flashy hydrograph (Figure 11), the replacement of natural surfaces by artificial ones has the opposite effect. Two characteristics of urbanisation impact basin response: firstly, the replacement of vegetated soils with less permeable surfaces such as tiled roofs, tarmac roads and concrete paving. Even today, when strict planning has prevented the unplanned expansion of towns and cities in the UK and Ireland, the replacement of permeable with impermeable surfaces continues. In-fill housing and the demands for off-road parking both contribute to the increased proportion of impermeable land cover (Figure 12). Secondly, urban environments are provided with highly engineered drainage systems. Buildings have guttering and downpipes, roads are cambered and have kerbside drains, while beneath the surface a complete system of pipes direct water efficiently into nearby rivers and streams. Even these natural routes themselves have been channelised and modified to ensure rapid drainage.

Figure 12: Paving over front gardens for off-road parking space in Putney, London

Fluvial processes

Rivers at work

Drainage basins not only involve the movement of water and sediment, but there are also transfers of energy through the system (Figure 2). Water on land has potential energy, which becomes kinetic or movement energy when it starts to flow down slopes. In river channels this energy is used for three types of work: friction, erosion and transport. Energy is lost by **friction** between the water and the bed and banks of the channel; this is most obvious where the channel has large boulders as often seen in mountain streams. **Erosion** is the wearing away of the bed and bank of the channel and **transport** is the carrying away of this eroded material. Material carried by a river is called the **load**.

Erosion

Most erosion by rivers takes place during high flow or flood periods when the volume and velocity is greater and the river energy is at a maximum. This is why some rivers seem unchanged over many years as flood conditions may be rare. Four different river erosion processes are identified, although all may happen simultaneously during times of high discharge.

1. **Abrasion/corrasion** occurs when rock fragments carried by the river wear down the bed and banks of the channel. Large boulders scrape the rock bed while sand and gravel can smooth the surfaces much like the action of sandpaper. In times of high discharge, the impact can be dramatic as bedload moves downstream.

2. **Hydraulic action** is the erosive power of the water itself. It is most effective on soft or loose (unconsolidated) bank materials and can often undercut the bank on the outer side of meanders. It can also weaken solid rock by forcing air into cracks, especially at waterfalls and rapids.

3. **Solution/corrosion** occurs when soluble rock or minerals are dissolved by the water. This is particularly effective on carbonate-based rocks (limestone and chalk) as river water is slightly acidic. Unlike the other processes, this form of erosion concerns chemical rather than physical change.

4. **Attrition** is the wearing down of the river load itself as particles strike each other and the bed and banks. The particles will then reduce in size and become more rounded as they travel downstream. Attrition, therefore, is a process impacting the river's load and not the shape of the river channel itself.

Erosion processes operate in two ways:

- Vertical – deepening valleys in mountains as abrasion and hydraulic action lower the channel bed producing a steep sided V-shaped profile.
- Lateral (horizontal) – meandering rivers create wider shallow valleys with floodplains.

Transport

Rivers gain their load either by the erosion processes or from material falling or washed in from the valley sides. The sediment load carried by a river can be described in three parts, related to transport: **solution load**, **suspended load** and **bedload**. Solution load is material dissolved in and carried by river water while suspended load is held up and carried along by the river, often discolouring the water. Figure 13 illustrates each of these, and the four processes involved in their movement downstream. The movement of bedload downstream is associated with high or very high river discharge and is described either as a rolling motion, termed **traction** or a bouncing, skipping action, termed **saltation**.

Why would solution be a more significant load transport process in regions with limestone rocks and/or tropical climates?

*Figure 13:
River transport
processes*

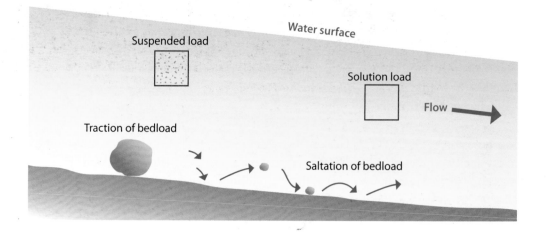

Deposition

Rivers need energy to overcome friction and to erode and transport their load but when a river loses energy the load can no longer be carried and so deposition occurs. River velocity is a good indication of river energy, so where and when rivers slow down deposition is likely. The most obvious place for a river to slow is where it reaches its mouth and flows into a lake or the sea. Here sediment falls to the bed, starting with the heaviest, largest particles (such as gravel and sand) while finer sediment (such as silts and clays) may be carried further. More detail of this process is given in the description of delta formation later in this chapter (page 23). Rivers also slow down along their course. A common site for deposition of load is on the inner bank of meanders, known as **point bar deposits**. As a river channel swings around, the water flowing on the inner bend has a shorter distance to travel and so it slows down. As a result, the processes of erosion and deposition can occur at the same time on opposite river banks only a few metres apart. Any material deposited by a river is termed **alluvial**. When a river slows down along its course it may start to deposit the sediment load on its bed and small islands called **eyots** may form. The river channel itself can become divided into several smaller interweaving channels in a process called **braiding** (Figure 14).

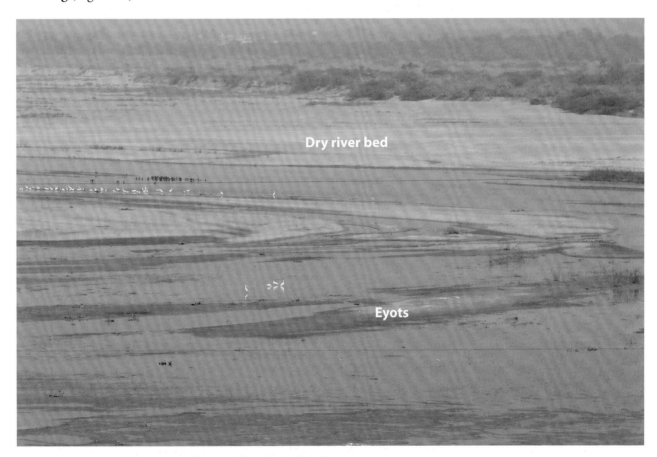

Figure 14: Braiding of the River Krishna, Andhra Pradesh, India

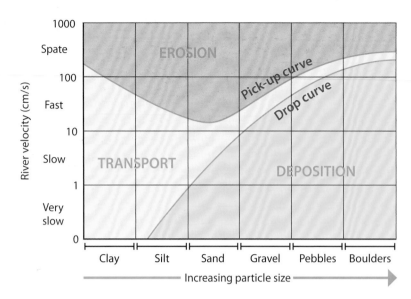

Figure 15:
Hjulström
curve

Hjulström curve

This shows the energy (velocity) needed to erode (lift), transport or deposit different sizes of river load.

Two key ideas are:

- The larger the particle, the more energy needed to erode or transport it.
- Erosion needs more energy than transport, ie once a particle is lifted it requires less energy to keep it moving.

Figure 15 illustrates these two concepts, where the vertical axis represents energy as river velocity and the horizontal axis is sediment size.

When real rivers and streams are studied, the actual relationship does not hold true for the smaller sediment. Hjulström's curve shows that small clay and silt particles require much more energy than expected to lift them up (erosion), although the energy needed to transport them is, as expected, low.

What is it about the small particles that cause this anomaly?

There appear to be two factors that cause the apparent anomaly in the Hjulström curve:

- Firstly, clay particles have a natural cohesion, they stick together so requiring more energy to lift (erode) them. Once separated, the individual clay particles are very easily carried and remain suspended in water, even at very low velocities.
- Secondly, clay particles can be tightly packed. Imagine a river bed covered in gravel; its surface will be rough and varied and the water flowing over it will be disrupted and turbulent. If the same bed is covered by a layer of clay particles, the flow will be smooth with little friction, so the clay is less likely to be disturbed by the water flow.

Exercise

1. A river has flooded but the discharge is now falling and the velocity of the river is slowing down. Using the Hjulström curve, explain what will happen to the river's load as it slows from a velocity of:

   i) 800 cm/s to 100 cm/s.

   ii) 100 cm/s to 10 cm/s.

   iii) 10 cm/s to 1 cm/s.

River landforms

Waterfalls

Perhaps the most dramatic and well known river landform is the waterfall. The name succinctly describes this feature, as it occurs when the water in a channel falls vertically down a rock face. There are several possible reasons for the initial formation of a waterfall, but their features and processes are similar. A common initiating feature for waterfalls is a band of resistant rock (cap rock) across which a river runs. Over time, as the river flows from the harder to the softer rock, it cuts down into the softer rock more rapidly, forming a steep section or step in the river bed. As the water starts to fall vertically, it quickly erodes the weaker rock and forms a plunge pool. Abrasion and hydraulic action processes combine to deepen and widen this pool, and also to undercut the hard rock above. Eventually, part of the hard cap rock collapses into the

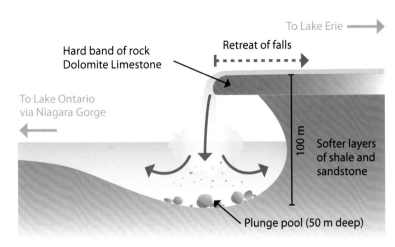

Figure 16: Sketch cross-section Niagara Falls

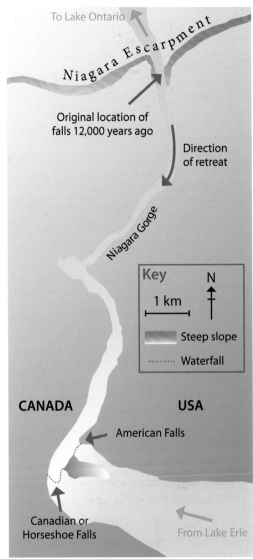

pool, providing more material for the abrasion process, and the position of the waterfall moves upstream. This gradual upstream progression of the waterfall is termed 'retreat' and creates a steep-sided gorge, clearly recording the former position of the fall.

Niagara Falls: The cap rock, or fallmaker as it is termed in North America, is a hard band of limestone. At the end of the last ice age, water from Lake Erie flowed northwards into Lake Ontario. At this time a 100 m high waterfall existed, which has, over the last 12,000 years, retreated south to become the two 50 m high Horseshoe and American waterfalls. This wandering retreat is marked by the 11 km long Niagara Gorge. In the future, the waterfall will continue to retreat and reduce in height until it reaches Lake Erie by which time it will have degenerated into a **cascade** (a series of small steps and rapids), a steep river section of white water flow.

Figure 17: Map of Niagara Falls

Figure 18: Detail of a fall, pool and gorge at Hardow Force, North Yorkshire

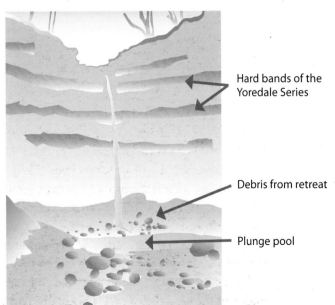

Meanders

If waterfalls are common on rivers then meanders are universal. Taking their name from a particularly sinuous, twisted river in Turkey, a meander is a river bend. Despite years of research there is still no agreed explanation for the formation of meanders in river channels. Many patterns have been noted, for example, meanders are more common in channels where the bedrock is neither too hard nor too fragile, the gradient downstream is gentle and the load carried is not excessive. Straight river channels are rare in nature but even in these, when the line of fastest water flow (called the **thalweg**) is plotted, it forms a twisted (sinuous) course (Figure 19). Meanders are not the result of obstacles in the river path but rather they seem to allow a river to balance its energy with its load. In the early twentieth century, river engineers shortened the length of the lower Mississippi to improve flood control and river transport. To do this, they cut through the river's enormous meander loops to create short-cuts, some saving 5 km on the old channel route. Within 15 years the Mississippi, through natural erosion and deposition processes, had restored most of its original length. If understanding why meanders occur is complex, it is much more straightforward to picture why, once a river bends, the meander will grow and develop.

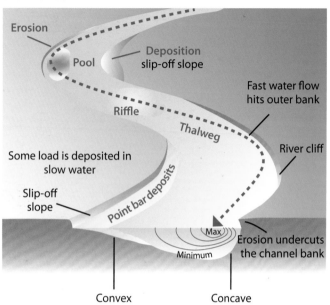

Figure 19: Diagram of a meander

Erosion

Deposition slip-off slope

Pool

Fast water flow hits outer bank

Riffle

Thalweg

River cliff

Some load is deposited in slow water

Slip-off slope

Point bar deposits

Max

Minimum

Erosion undercuts the channel bank

Convex

Concave

Riffles and pools – meander formation

One pattern closely associated with meanders is the sequence of regularly spaced **riffles** and **pools** along a channel. Riffles are sections of channel between two meanders where the water is shallow and flows through coarse gravel bed sediment. Pools are found in the channel bed near the outer bank of meanders (Figure 19); they are areas of deep, smooth water flow. Friction at the shallow riffles deflects the river flow towards one bank and so causes that bank to be undercut by erosion processes. In this way a meander is initiated downstream from the riffles and the channel erodes to match the swinging nature of the river flow. In reality, the thalweg is only one line of flow in the river; other flows form a helical or downstream corkscrew pattern.

Meanwhile, as a meander's outer bank is subject to erosion and undercutting on the opposite inner bank, water moves more slowly and material is deposited. In this way the river channel does not get wider but it moves sideways (laterally). Figure 20 shows how the riffles and pools created by the line of fastest flow in straight line channels remain as the river meanders widen and grow by lateral erosion processes. Meanders in cross-section have an asymmetrical appearance with a steep, often concave, outer bank or **river cliff** and a gentle inner bank called a **slip-off slope** on which **point bar** deposits are found (Figures 19 and 21).

Figure 20: Possible sequence in the evolution of a meandering channel

Key

– riffle

– pool

e – erosion at outer bank

Throughout the development from straight to meandering channels the riffles remain anchored in more or less a straight line.

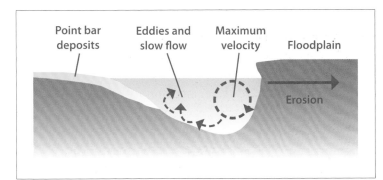

Figure 21: The asymmetrical channel cross-section of a meander

Figure 22: Meander features on the River Ure, Upper Wensleydale

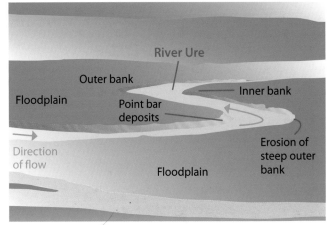

Meandering meanders

The overall width of a river's meanders is related to its load, slope and discharge, and their lateral growth is limited. However, meanders are not fixed but tend to migrate down the valley, just like a wave moving along a piece of string. The reason for this goes back to the pattern of outer bank erosion. The maximum erosion point is normally a little downstream of the mid-point of the bend due to centrifugal forces, and so the meander channel creeps downstream. The sketch map of the Mississippi River channel (Figure 23) shows how, in over 80 years, a meander grew laterally by over 1 km and migrated downstream by over 2 km. This illustrates one good reason for not using natural streams as political boundaries!

Oxbow lakes

Oxbow lakes, also known as **mort lakes** or **billabongs** (Australia), are the remnant of a former meander that has been naturally by-passed by the river channel. The usual sequence involves a meander becoming increasingly sinuous and its 'neck' narrowing to a short distance. Then, often in flood, the river cuts through the swan's neck and forms a new channel. For a time water flows across

Figure 23: Migration of Mississippi River meanders downstream

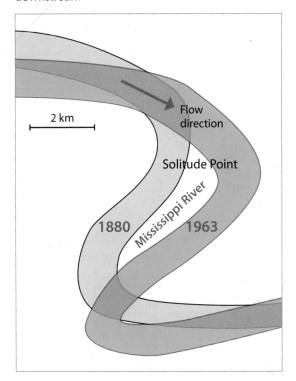

the new route and through the old meander but gradually deposition blocks off the meander, leaving it disconnected from the channel and creating an oxbow lake (Figures 24 and 25). The name comes from the similarity in shape to the U-shaped piece of wood of the same name fitted under the neck of a harnessed ox. Over time plant invasion and sediment washed into oxbows causes them to dry out. From an aerial view, their shape often remains visible as a **meander scroll** or **scar**.

Figure 24: Oxbow lake on the River Ure floodplain

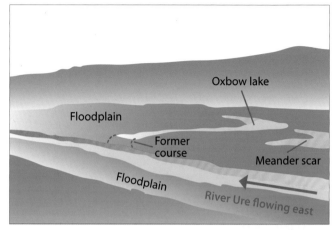

Figure 25: The formation process of oxbow lakes

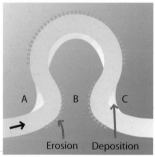
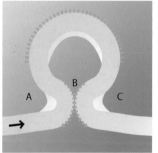
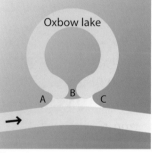

Step 1:
The meandering river deposits material on the slip-off slopes of the inner bends at A and C. Erosion on the outer banks forms river cliffs and the neck of the meander at B narrows.

Step 2:
Erosion, possibly during a flood episode, breaches the neck of the meander and some of the river's discharge follows the new direct channel downstream.

Step 3:
The meander loop becomes cut off from the new channel and the oxbow lake is formed.

Step 4:
Over time the oxbow lake will gradually fill in with sediment and plant growth to leave a meander scar on the floodplain (see Figure 24).

Floodplains and levées

Floodplains are most common where rivers have left the mountain stage of their course. As the name suggests, floodplains are flat areas adjacent to the river channel. They are normally covered by alluvial material deposited by the river itself but they are not merely deposition features. As noted earlier, river meanders migrate laterally and downstream so over time they gradually erode their way across a broad belt of land. This area is partly covered by point-bar deposits from the inner bank of the meanders but also with sediment (clay, sand and silt) deposited during flood periods. Thus both the processes of lateral erosion and deposition are involved in the creation of a river's floodplain.

In times of high rainfall or snowmelt, rivers sometimes exceed their channel's carrying capacity, forcing some water out onto the floodplain. This is known as **flooding** or **inundation**. Naturally, the water slows down as soon as it leaves the channel, losing energy and depositing some of the river's load. As the Hjulström curve shows, the largest gravel or sand-sized particles settle first, followed by the silt-sized particles, which are deposited further away. Finally, the finest clays are often carried far from the channel, settling as the water infiltrates into the soil or evaporates. Such floodplains often have a convex upward profile.

One distinctive feature of such floodplains is the natural bank of coarse deposits alongside the channel itself. These natural raised river banks were first named by French settlers along the Mississippi river as **levées**. Often these features of natural deposition are reinforced and raised to provide flood protection. It was one such levée on the Mississippi that tragically broke in 2005 during Hurricane Katrina, causing the devastating flooding of New Orleans, sited on the floodplain. Natural levées can cause difficulty for tributary streams trying to break through to join the main channel. Often such streams flow parallel to the channel for miles downstream before finding a break in the levée. These are technically called **deferred tributaries** or, more memorably, a **yazoo** after a Mississippi tributary of that name (Figure 26).

Figure 26: Sediment deposition on the floodplain

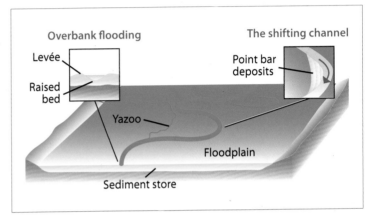

Deltas

Deltas are areas of deposited alluvial material, which occur when the river reaches its mouth. They are formed where a flowing river meets a sea or lake, and slows, depositing material faster than the sea can remove (erode) it.

Deltas can be found on many rivers, including some of the world's largest, such as the Ganges-Brahmaputra, Amazon, Colorado and Nile. Their formation is affected by a number of factors:

- Sediment load – Deltas are common at the mouths of rivers with large sediment loads, such as the Nile.

- Coastal energy – Deltas are common in low energy coastal environments, such as the Mediterranean, where river sediment gradually builds up on the sea-bed without being disturbed (deposition exceeds erosion). They are rare along the coast of the UK and Ireland due to the large tidal ranges and high energy environments, which erode the coastline (erosion exceeds deposition) and prevent deposition.

- Flocculation – Deposition at the coast is aided by the process of flocculation. This is a chemical process in which the salt nature of sea water causes fine clay particles to join together and allows them to settle. Remember that by the time a river reaches the sea most of its load will be composed of finer particles (sand, silt and clay) due to attrition.

- Sea floor gradient – A gentle and shallow sea floor gradient allows sediment to build upwards to reach the surface as a landform.

Deltas are often regarded as an extension of the river's floodplain, extending the coastline out into the sea. A common feature is that the river itself breaks up i many channels across the delta, causing braiding (Figure 13). These separat are called **distributaries** as they effectively **distribute water and sedime** river numerous 'mouths'.

Three different types of deposit are usually found in deltas. As illustrated in Figure 27, these are the topset, foreset and bottomset beds. The lightest and smallest particles of the river load are carried out into deeper water to settle as bottomset beds. The medium sized particles drop as steep wedges of sediment (foreset) and the heaviest particles settle rapidly on the top of the delta (topset). Over time, as the delta extends seawards, the bottomset and foreset beds are further buried and incorporated into the delta itself.

Figure 27: The three sediment beds of a delta

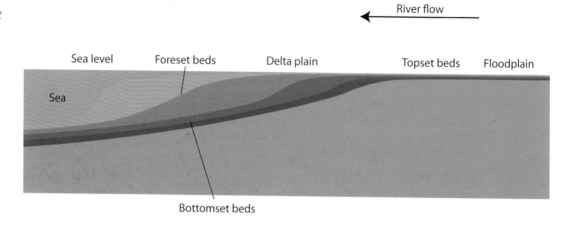

Deltas are usually classified by their shape, which is largely the result of how the river load is deposited. Based on their appearance, the two most common delta forms are the **Arcuate** and **Bird's Foot** deltas (Figure 28).

Figure 28: Common delta forms

Arcuate form

The River Nile's delta gives all deltas their name as it is shaped like the Greek letter delta (Δ). This is the **arcuate** form, with a convex, curved outer edge maintained by the action of longshore drift along its outer sea-edge. The Nile delta is readily identified on satellite images as it forms an area of well-watered agricultural land in this desert country. The Nile's delta sediments are several kilometres deep, having been deposited over thousands of years of annual river flooding. However, the control of the river by the High Aswan Dam in the last 50 years has meant that less sediment is carried by the Nile to its mouth in the Mediterranean Sea. Today, the sea is invading Egypt's delta lands along the north coast, threatening this invaluable farmland.

Bird's Foot form

The Mississippi River drains almost 50% of the USA and carries a huge sediment load into the shallow, low-energy waters of the Gulf of Mexico. Here the river flows out into the Gulf, depositing sediment along its channel sides. Additional distributary channels also extend, and so the distinctive bird's foot shape is created and recreated over time.

River deltas not only create new land, they are usually highly fertile with an abundant water supply. It is no wonder that some of the world's most densely populated areas are on deltas including in North Africa the Nile, in South East Asia the Mekong and in Western Europe the Rhine.

Exam Questions

1. Study Resource A which shows the annual rainfall and runoff for the River Thames at Abingdon.

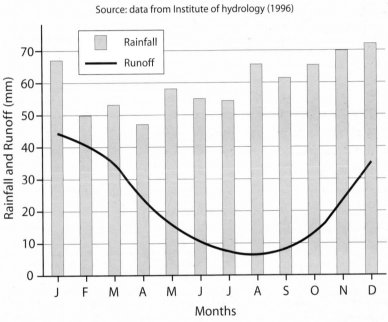

Resource A

Source: data from Institute of hydrology (1996)

(i) In which months is rainfall at its highest and lowest? State the rainfall amount. [2]

(ii) Describe and explain the changes that occur in runoff throughout the year. [4]

Question from CCEA AS1 Physical Geography June 2010 © CCEA 2016

2. Explain, using annotated diagrams, the river processes involved in the formation of natural river levées and deltas. [12]

References

Geofile articles:

'Hydrographs: Physical and human impacts', *Geofile* 542, series 25, 2006–2007

Geo Factsheets:

'Storm Hydrographs', *Geo Factsheet* 83, Curriculum Press

'River long and cross-valley profiles', *Geo Factsheet* 215, Curriculum Press

Case studies

LEDC flooding event: Pakistan

MEDC flooding event: Somerset Levels

Students should be able to:

(i) explain why some rivers need to be channelised and how this is achieved through realignment, resectioning and dredging

(ii) understand how and why environmentally sensitive and sustainable management strategies are needed to manage river channels

(iii) investigate the causes of recent flooding and its effects on people, property and the land

Rivers are simultaneously a vital resource for people and a potential lethal threat. For both reasons, around the world we have interfered with river channels over a long time. Improving technology, from the spade to the JCB, has allowed the scale and scope of such interference to grow, usually to our benefit but often too with unexpected and undesirable consequences.

The need for channelisation

Channelisation is a term that describes the deliberate modification of natural river channels. These involve changes to the width, depth or the plan form (vertical view) of the channel and also diversions of channels and the construction of embankments or artificial levées (Figure 29).

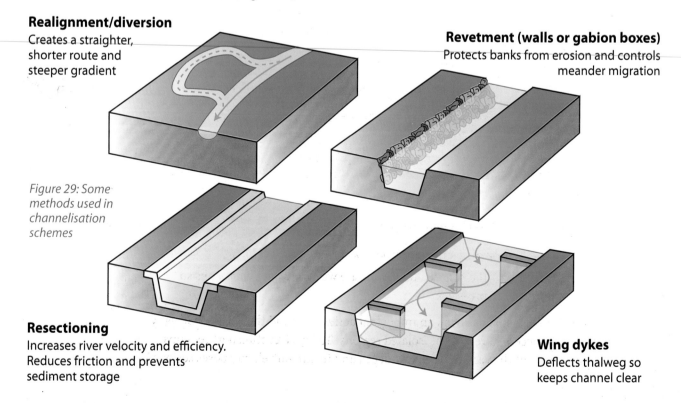

Realignment/diversion
Creates a straighter, shorter route and steeper gradient

Revetment (walls or gabion boxes)
Protects banks from erosion and controls meander migration

Figure 29: Some methods used in channelisation schemes

Resectioning
Increases river velocity and efficiency. Reduces friction and prevents sediment storage

Wing dykes
Deflects thalweg so keeps channel clear

An analogy – river discharge and road traffic

The problems with river channelisation are similar to the issues facing planners attempting to improve traffic flow along roads. Imagine a commuter's drive to work. Along the journey there are four major junctions. At each of these junctions traffic builds up during the rush-hour. Planners then place a roundabout at the first junction and traffic flows more freely. Unfortunately this simply speeds more traffic to the second junction, where even longer tailbacks develop. Planners act again, placing a phased set of traffic lights at the second junction. This only results in sending even more traffic, more quickly, to the third junction causing longer queues and increased congestion.

Instability

Natural river channels are constantly adjusting their shape and form through processes of erosion and sediment deposition. Engineering schemes can immediately and radically change the channel characteristics. The response to this rapid and unnatural change often leads to instability. Change in the channel shape along one reach of a river not only alters its flow characteristics, it also creates a new regime further downstream beyond the section altered. The Mississippi navigation scheme (Figure 31) illustrates an example of instability following channelisation.

One of the oldest forms of channelisation involves the raising or creation of levées. Natural levées form when rivers flood and deposit coarse load immediately on their banks. These can be enhanced locally to improve the channel cross-section and thereby reduce the flood risk. However, levées alter how rivers move their sediment load. Firstly, reduced flooding means less alluvium is spread across the floodplain. Secondly, the load starts to be deposited on the river bed itself, raising the river channel above the floodplain level. This not only makes it difficult for water on the floodplain to drain into the river but should floods occur, their impact can be more severe. Embankments along the River Ganges, as it flows through the Bangladesh capital Dhaka, were designed to reduce the seasonal monsoon floods. In reality, the normal four to six weeks flooding of low-lying areas of the city was extended to four months. This was because monsoon rainfall that would previously have drained into the river was ponded back by the new embankments and could only drain away when river levels fell.

Increased flood hazard

Moving a greater volume of water more quickly to an unchanged section of river channel is inevitably going to increase the risk of that river exceeding its bankfull discharge and inundating its floodplain. Some observers suggest that the Mississippi floods of 1993 and 2010 did not happen in spite of 100 years of management but rather because of it. In many countries attempts to reduce flooding of urban areas through resectioning, realignment and containing rivers ultimately creates a flood hazard downstream where the engineering of the channel ends. This is because the nature of surfaces and drainage systems in urban areas shed water rapidly and the newly modified channels accelerate discharge out of town on into the unaltered river channel with its lower carrying capacity.

Aesthetic consideration

River courses are often highly valued as an amenity by the local community. This ranges from those who use the river for sport, such as angling or canoeing, to those who walk along the banks, or others who view the river as an element of the scenic environment. The most common response to channelisation is an objection to the canal-like appearance and the intrusive hard engineering structure built. In the words of David Shaw, an angler on the River Main before but not after its channelisation, "I have no interest in sitting for an evening's fishing at a glorified drainage ditch that was once a beautiful stretch of winding river" (Figure 34). Channelisation often results in the loss of the stream's riffles and pools and also produces a wider variation in discharge, with low flow levels which expose the bed becoming more common.

Figure 34: Map and photographs of the River Main scheme

Modern sluice gates

View north from bridge

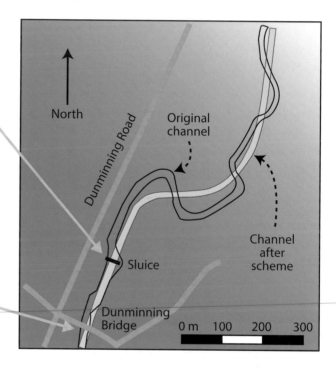

Impacts to river ecology

Rivers provide a diversity of habitats and environments that interact with the biotic and abiotic elements of ecosystems in a complex way. Shallows or **riffles** and deeper sections called **pools** are natural elements of river meanders. These provide a range of habitats for both plant and animal life. Coarse gravel beds are used by fish laying their eggs, while pools provide shelter for resting fish, especially in times of low flow and where these are shaded by overhanging vegetation. Figure 35 shows how channelisation can impact river bed and bank environments. Land alongside rivers is known as the riparian environment and this zone will be directly impacted under many engineering schemes. For instance, improved floodplain drainage makes it drier, so the flora and fauna of marshes and wetland suffer. Channelisation usually modifies river banks either through the regrading of their profile or the clearance of vegetation.

Bank vegetation provides food and shelter for insects, mammals and birds that make use of the river. Plant cover on river banks reduces erosion, as roots bind the soil particles and plants shade the river, helping to moderate changes in stream temperature. This latter point is critical to many plants and animals that are sensitive to wide fluctuation in light and temperature conditions. Increasing concern for the ecological impacts of channelisation, as well as fears over its long-term effectiveness, has led to a rethink in channel management. In Northern Ireland, drainage schemes such as the River Main used detailed public enquiries and environmental assessments as, gradually, a new school of thought emerged. This modern approach focuses on the creation of environmentally sensitive and sustainable solutions, incorporating hard and soft engineering, and river restoration.

Figure 35: The impact of channelisation on river bed and bank environments

Natural channel

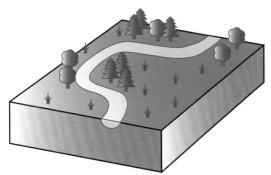

Suitable water temperatures; shading; good cover for fish life; abundant leaf input.

After channelisation

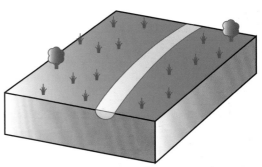

Increased water temperatures; no shading; no cover for fish life; rapid fluctuations in temperatures; reduced leaf input.

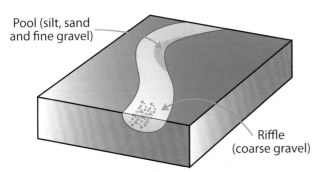

Pool (silt, sand and fine gravel)

Riffle (coarse gravel)

Pool-riffle sequence; sorted gravels provide diversified habitats for many stream organisms.

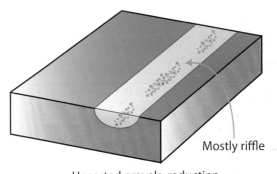

Mostly riffle

Unsorted gravels; reduction in habitat variation.

The pool environment

Bankfull discharge

Disperse water velocities; high in pools, lower on riffles.

Bankfull discharge

Increased stream velocities.

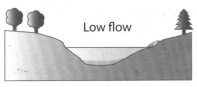

Low flow

Moderate depth in dry seasons.

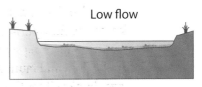

Low flow

Much reduced depth of flow in dry seasons.

Figure 36: The River Main Drainage Scheme Public Enquiry

This engineering scheme was designed to improve the land drainage for 4,000 hectares along a 20 km section of the River Main in County Antrim. The river rises near Ballymoney and flows south past Ballymena into Lough Neagh. The proposal involved the enlargement, regrading and realignment of the river channel, and the replacement of an existing weir with sluice gates (Figure 34). The enquiry considered the views of all interest groups including:

- several local and national representatives of angling clubs.
- many local farmers and representatives of farming organisations.
- owners of industrial premises that used water from the Main for industrial processes.
- numerous conservation and heritage groups and experts.

A large part of the report and its conclusions dealt with the need to modify plans in light of their potential ecological impact, especially of fisheries, riparian ecosystems and scenic amenity. Changes included:

- restoring pools and replacing boulders to maintain fish shelter.
- working from alternative banks to leave vegetation untouched.
- retaining some river bends for habitat variety.
- considering the construction of a fish pass at the sluice gates.
- increasing the number and scale of replacement trees on the banks.
- the relocation of the channel to avoid a particularly scenic landscape at Gracehill.

Creating environmentally sensitive and sustainable management strategies

In reassessing the engineering and management of rivers it has been suggested that planners need to move away from the view of 'rivers as hazards and wealth' to 'rivers as resource'. Uses such as recreation, conservation, aesthetics and ecology need to be fully valued. The damaging impacts and expense of some channelisation projects encouraged the development of soft engineering methods to replace or mitigate hard engineered schemes. Hard engineering refers to the structural approach to issues. In the case of rivers these include dams, revetments, land drainage systems, culverts, sluice gates and artificial levées. Soft engineering is usually defined as solutions that work with nature rather than against it. In reality many projects incorporate elements of both hard and soft engineering. Balancing the demands of the various groups interested in rivers and basins, while planning for the future, is a difficult task. Soft engineering techniques and 'softer' structural approaches can help achieve the goal of environmentally sensitive and sustainable solutions.

Soft engineering techniques

Two examples of these techniques used in cases where river flooding is a hazard, are **afforestation** and **land zoning**. Planting trees or other vegetation across catchment areas has the dual effect of firstly, reducing the amount of rainfall that leaves the basin by river channels and secondly, slowing down the transfer of water through the basin into the river. Land zoning involves the management of the land use to reduce the risk or scale of flooding. In places wetlands may be retained as water storage areas or low-lying land

Figure 37:
Photographs of flood
defences on River Tay
in Dundee

that has not been developed may be set aside to temporarily store floodwater as required. In Dundee, on the banks of the River Tay, sports fields and public parks line the river bank both up and downstream of the city centre. During high water these can be flooded without causing great expense or long-term damage. In the town centre a more structural flood defence system is employed, including steel flood barriers across river access points and roads, as illustrated by the photographs in Figure 37.

Softer channel engineering techniques

One approach is to modify the traditional resection approach by using partial channel dredging. Natural banks are retained while part of the bed is lowered to increase the channel capacity. This allows most of the ecological habitats to remain intact though it is not clear if it does help reduce flooding effectively. Another method has been to pull back from the channel itself to leave a river corridor along which the river is free to flow and even modify its own path and channel. Defences are then placed some distance from the edge of the river channel. Embankments along the floodplain rather than on the river banks mean that flooding can occur but only to a fixed point. This additional storage land may then protect land further downstream from inundation.

The provision of an additional channel running parallel to the existing river but only used during high, overbank flows, is an example of a hard engineering structure used in a soft way. The river is left untouched with all its diversity of ecological habitats intact. Biological engineering is a term sometimes applied to the use of specific types of planting. Green stakes of willow can be used to line river banks as a living revetment or woven willow cages used as gabions to stabilise banks at bends or bridges. These features encourage the trapping of sediment, allowing river plants to establish and flourish in and near the water's edge.

River restoration is the application of more environmentally sensitive methods to rivers that have previously undergone hard engineering channelisation. Over the last 20 years across the UK and Ireland, many attempts have been made to recreate the more natural appearance of modified rivers in both urban and rural settings. Common changes are the reinstatement of bends and meanders, the restoration of pools and riffles on the bed, and varying the cross-section channel shape from the efficient but monotonous trapezoid plan. River banks can be vegetated and regraded to improve access for anglers, walkers and other recreational users. River restoration is usually applied to short sections of rivers (Figure 38).

Figure 38: The East Belfast Flood Alleviation Scheme Orangefield Park section, Knock River

In East Belfast the Connswater Community Greenway is a £32 million scheme to create a 9 km linear park following the Knock, Loop and Connswater Rivers. One element of the project is the East Belfast Flood Alleviation Plan, delivered jointly by Belfast City Council and the Rivers Agency. Areas within the drainage basin of the three channels are under the threat of flooding by the rivers or the sea. This risk to people and property became apparent after exceptional rainfall and flooding in June 2012, and again following coastal storms in January 2014. The scheme provides enhanced protection for some 1,700 properties.

Phase 1 of the larger scheme was completed in 2015. It included the re-routing of the Knock River through Orangefield Park in a residential area of East Belfast. The aims were to provide improved flood protection to nearby homes and to develop the river as a key environmental feature of the public park amenity.

The scheme started with the removal of the existing hard engineered concrete channel. The new plan incorporated environmental restoration with a wider, winding, more natural channel created for the Knock River. In addition, wetland habitats were designed adjacent and linked to the new channel. A new pedestrian bridge now straddles this realigned channel, along with the grassland and wetland ponds that are designed to flood and temporarily store flood water during high flow periods.

Through the park a mixture of hard engineering structures (flood walls and culverts) and soft engineering systems (wet wildflower meadows and ponds) have been employed in this river restoration scheme. The winding path that follows the river route through the park has been named the 'Marshwiggle Way' after a character in the Narnia chronicles written by local author CS Lewis.

The Knock River in Orangefield Park: before and after images of the river restoration program (2014–2015).

Source: http://www.connswatergreenway.co.uk/section-c3/knock-river-section-2-ccg-interface-to-grand-parade

Artificial shallows (riffles) and wildflower planting are designed to restore plant and animal habitats along the channel.

Pedestrian access to the park by bridge is possible even when the river overflows its channel to incorporate the designed wetland pond habitat in the centre.

New culvert on the Knock River October 2014 and September 2015.

CASE STUDY: Pakistan 2010, flooding event in a LEDC
Background

Figure 39: Location of Pakistan

Much of the Asian nation of Pakistan has a hot tropical desert environment, with annual rainfall totals below 250 mm in the central and southern regions of the Punjab, Sindh and Baluchistan. Only in the north is rainfall sufficient to feed the headwaters of the nation's major river drainage basin, the Indus.

The people of Pakistan live with a climatic reality that they share with around one third of the world's population; it is known as the monsoon climate. While monsoon literally means 'season', the implication is that the annual climate is divided between a dry and warm to cool winter monsoon and a hot wet summer monsoon. These summer monsoonal rains are the agricultural lifeblood to many nations, including Pakistan. Not only is summer rain and subsequent river flooding normal in Pakistan, it is regarded as essential to renewing groundwater supplies, supporting aquaculture and fish farming, and providing adequate irrigation for the rice and wheat crops. Pakistan has a short and often unreliable wet monsoon that starts in late June and lasts for around three months. As a predominately rural and therefore agricultural based national economy, these rains are critical to the survival and developmental progress of its people.

Figure 40: Impacts of the Pakistan floods of 2010

Source: Data from *Financial Times*

Impact on people:
Deaths: 1,600
Injuries: 2,366
Affected: 17.2 million

Impact on schools:
Damaged: 10,916
Used as shelters: 6,097

Impact on agriculture:
Crops destroyed: 3.6 million hectares
Livestock killed: 1.2 million
Poultry killed: 6.0 million

UN emergency funding:
US$168 million
(approx £130 million)

Causes
Physical

The summer rainfall of the wet monsoon arrives after warm air over Cental Asia rises, drawing in moist air from the Indian Ocean to the south east. The north westerly winds travel across neighbouring India and bring rains to the northern part of the country. Pakistan has the latest arrival of these summer rains and is also the first region on the sub-continent from which they withdraw.

The 2010 summer monsoon brought exceptionally heavy rainfall, with most areas receiving twice their average rainfall in July and August. The nature of the rainfall in short but intense downpours meant that the Indus river system was regularly overwhelmed so flooding its extensive floodplain. While some annual flooding was expected in 2010, the inundation was the worst for over 80 years.

In South Asia, the summer monsoon is studied intensely as it is critical to the region's economy and society. In 2010, unusual flows in the upper atmosphere blocked the normal progress of the jet stream that guides the rains, including very heavy rainfall over the Western Himalayan mountains. This created a deluge of water into the upper tributaries of the Ganges.

Flooding 30 Aug 2010:
■ Severe
■ Moderate

Swat Valley

Disputed (controlled by Pakistan)

Khyber Pakhtunkhwa Province

Islamabad

Disputed (controlled by India)

Indus

Lahore

Punjab Province

0   200 km

Baluchistan Province

Indus

Sindh Province

Hyderabad

Karachi

Figure 41: Inundation of agricultural land in the Sindh, Pakistan 2010

Source: Photo by Staff Sgt Wayne Gray, Pakistan Humanitarian Aid Flood Relief, (Flickr creative commons license) http://www.flickr.com/photos/dvids/4996560018/

Many meteorologists have linked the unusual pattern to the 2010–2011 La Niña cycle. This is an occasional reversal of ocean currents and winds in the Pacific that is regularly linked to extreme meteorological events (see page 120), including in this case tornadoes in the USA and floods in Queensland, Australia.

Human

It is clear that some human activities in Pakistan have contributed to the increased risk of flooding. During high flow, the River Indus carries a large load of silt, especially on its bed. This reduces the channel's carrying capacity and frequently river embankments have been raised to ensure that flood water can be contained. In 2010, the exceptionally high river discharge overtopped these raised embankments. The flood water that spread over the countryside was then trapped, as the embankments prevented water from draining back to the river channel when discharge levels fell. In places the summer flooding of 2010 did not retreat until 2011.

For centuries the Indus river has been the source of water for irrigation, especially in the dry winter season. For this reason Pakistan has thousands of kilometres of irrigation barrages, canals and channels. The routine maintenance of this system has been poor in places, with channels choked by silt which then fail to contain water flows.

With a population of 191 million people, the sixth largest in world, and increasing by 3 million people each year, the pressure for food production and house building has lead to forest clearance and the expansion of urban areas, with their impermeable surfaces causing rapid surface runoff. Pakistan's deforestation rate of 2% per annum is one of the world's highest and the nation has lost a quarter of its forest cover in the last 20 years. In the Swat Valley of North Pakistan, deforestation, initially by Taliban militants and recently by the timber businesses and rich landowners, has increased the flood risk. The loss of dense woodland increased soil and river bank erosion thereby decreasing the capacity of river channels during the heavy rains.

Impacts

This monsoon flooding claimed the lives of over 1,600 people, injured almost 2,400 others and directly affected up to 20 million people across Pakistan. The floods put millions of people out of their homes, and caused destruction to houses, roads, schools and health facilities. Agricultural production was heavily impacted with losses to crops, livestock, infrastructure and equipment.

People

About 1.6 million houses were damaged or destroyed; most were in the densely populated Punjab state. The flooding also caused extensive damage to key social amenities, including schools and health centres. According to UNICEF 7,600 schools need to be rebuilt while 436 health facilities were damaged or destroyed, so limiting health care services to local communities.

Two weeks after the initial floods over 10 million people were homeless and in critical need of food and clean water supplies. Ironically, the floods threatened domestic and commercial water supplies as most ground water wells were unusable, blocked by mud and silt. Stagnant flood water is an ideal breeding area for the malarial mosquito and this, along with the threat of water borne diseases such as dysentery, diarrhoea and cholera, was an urgent concern for the government and public alike.

The communities were impacted by the extensive damage to communication routes and infrastructure, with remoter villages isolated for weeks (see 'Property'). The loss of food production and the transport difficulties disrupting trade increased average local food costs by between 15 and 25%.

Over half of families in the flood-affected provinces of the Punjab and the Sindh depended on crop or livestock farming as their primary income (see 'Property'). This meant that most displaced people had little or no way of earning money in the short-term.

Figure 42: Long-term flooding disrupts many aspects of everyday life in Pakistan, including the agricultural economy and education

Source: Department for International Development

Property

As previously mentioned, 1.6 million houses were damaged or destroyed and there was extensive damage to social amenities.

Damage to the region's road infrastructure was extensive, with over 25,000 km damaged. This was especially apparent in mountain areas where many bridges collapsed cutting communities off from essential supplies. For example, all bridges along a 140 km stretch of the River Swat were destroyed. The flood also damaged phone and electricity supply lines, interrupting these services in many large towns.

Concerning agriculture, primary infrastructure such as tube wells, animal sheds, personal seed stocks, fertilisers and agricultural machinery were damaged or destroyed. According to an agricultural assessment report by the Food and Agriculture Organisation (FAO), the monsoon floods caused unprecedented damage to arable and pastoral farming as well and other primary economic activities, namely fishing and forestry. The floods came immediately before the main harvest of key crops, including cotton, rice, maize and sugarcane, and in advance of the planting of the winter wheat season (Rabi) which normally starts in September/October. Production loss of sugar cane, rice and cotton in the Punjab and the Sindh was estimated at 13.3 million tonnes (FAO). Two million hectares of standing crops were lost or damaged and over 1.2 million head of livestock (not including poultry) died due to the flood. In the aftermath disease and fodder shortages threatened 14 million livestock.

Early estimates on the cost of rebuilding the lost and damaged infrastructure and social amenities were around £7 billion.

Land

Natural forests and recently established community forest plantations were damaged by fast flowing flood water, leaving these lands unstable and susceptible to further soil erosion.

The floods triggered mudslides on steep mountain slopes in northern provinces and destroyed breeding grounds for wading birds. On tributaries of the Indus River an estimated 89% of habitats of reptiles and small mammals were affected. The floods swept large quantities of petrol and diesel into the river system polluting the water downstream. Health hazards linked to this pollution and with the large areas of standing water on the floodplain were monitored carefully for months after the event. Wildlife including hog deer and bears were killed in significant numbers by the flood and even the Indus River dolphin population was reduced.

As the flood waters did not recede for months in parts of the Sindh Province, farmers were unable to plant wheat for the Rabi season, a crop that normally depends on irrigation water. In the more mountainous provinces of northern Pakistan the flood waters receded more readily and overall crop losses were lower, 45% in the Khyber Pakhtunkhwa (KPK) compared to 60% in the Sindh.

The regular annual floods in Pakistan help to replenish the groundwater store for winter supplies for wells and pumps and this exceptional flood did revive the mangrove ecosystem of the Indus Delta, which had been in decline for over 50 years. However, these are the rare positive outcomes of the 2010 monsoon flood.

"The floods of 2010 are being termed as SUPER FLOODS for Pakistan due to their large-scale devastation in the country. The floods affected 21% of all cultivable land and uprooted 20 million people from their homes and lands. It was important to help people get back to their homes quickly to avert another food disaster in the country..."

Waqas Hanif – Advisor National Disaster Management Authority (NDMA)

CASE STUDY: Somerset Levels, England 2014, Flooding event in a MEDC

Background

Severe Atlantic storms during the winter of 2013–2014 caused significant flooding in the south of England and Wales. The Met Office described the December to January period as the wettest in over 130 years. One region that suffered its worst flooding for a century is the low-lying wetland district known as the Somerset Levels. Here prolonged flooding isolated villages and individual homes for weeks and in some cases for months.

The Somerset Levels is a unique environment where human activity and the natural world has developed a distinctive, wildlife-rich landscape. The Levels form a wetland coastal plain lying between the Mendip and Quantock Hills in South West England (Figure 43). They cover an area of 650 km² most of which is only 3–4 metres above sea level. The land is drained by several small rivers including the Parrett, Axe and Tone. Near the coast the underlying geology is clay while further inland peatland dominates. A millennium ago the region was a largely unoccupied marshland with a few isolated islands where monks lived seeking isolation. It was these religious communities, such as that at Glastonbury, that first initiated the reclamation of the marshland for farming. For over 800 years people have been trying to drain and reclaim the land for agricultural use and today 70% of the levels are under grass for grazing and the rest is used for arable farming. Willow is grown commercially and peat is extracted from the area. Today, artificial river courses, pumping stations and an extensive drainage system, including ditches known locally as rhynes, are used to remove excess water and maintain the local economy.

Figure 43: Map showing the Somerset Levels and the surrounding area. The major hill ranges and rivers are shown.

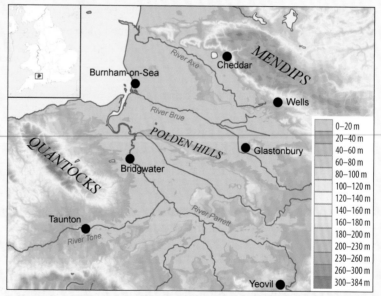

Source: Contains OS data © Crown copyright and database right (2016) used under the terms of the Open Government License. Map by Nilfanion.

Causes
Physical

The obvious cause of these floods was the unusually wet weather during December and January. Over a nine week period a succession of storms battered the UK and, as the table in Figure 44 shows, rainfall totals were well above average across the UK but especially so in South and Central England.

2A Global biomes

Students should be able to:

(i) identify the global distribution of biomes – tundra, tropical rainforest, hot desert and temperate grassland

(ii) demonstrate knowledge and understanding of the climate and soils associated with tundra and temperate grassland biomes

(iii) evaluate the actual and potential impacts of climate change on tundra ecosystems

The living world is sometimes called the biosphere, which is confined to a narrow zone on the surface of the earth and within the oceans. Geographically, life can be divided into different regions, which contain distinct combinations of plants and animals. These are often described by their vegetation, such as the tropical rain**forest** or the temperate **grass**lands. These zones cover large geographical areas and are known as **biomes**.

Globally, the simplest classification of biomes is into land (terrestrial) and water based biomes. The two water biomes, marine (sea) and freshwater, cover most of the earth's surface but the number of different land biomes is debated. The six basic land biomes are: the tundra, taiga (coniferous forest), temperate (deciduous) forest, grassland, desert and tropical rainforest. Commonly, the grasslands and desert are further sub-divided into the tropical (savanna) and temperate grasslands, and the hot (tropical) and temperate deserts.

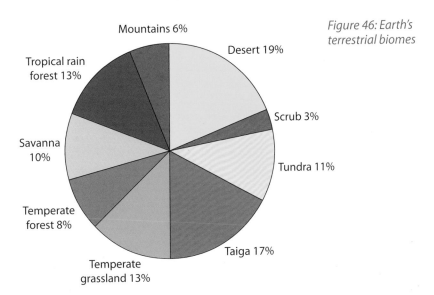

Figure 46: Earth's terrestrial biomes

The global distribution of biomes

Climate

Maps showing the distribution of the world's land-based biomes reveal that they tend to lie along lines of latitude. This is due to the fact that the dominant controlling factor is climate. If an area has a dry desert climate, the plants that grow there and the animals that depend on them will be adapted to these arid conditions. In hot deserts in different parts of the world from say, Australia to Southern Africa, plants with similar adaptations can be found. The species and family of plants and animals may be different but their nature and life patterns are often similar. Likewise, the native oak and beech woods of Western Europe are very similar in appearance to the mixed deciduous woodlands of the Eastern Coast of the USA, where maple and hickory trees flourish under a similar climate.

Figure 47: The impact of latitude on temperature variation

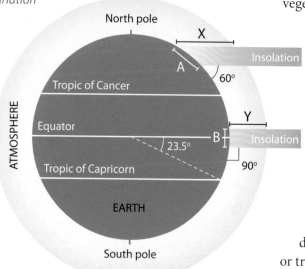

X and Y – Rays pass through different depths of atmosphere.
A and B – Same quantities of insolation heat different size areas.

There are several climate factors that determine the vegetation of an area:

1. Latitude

This mainly influences temperature and its seasonal variation (Figures 47 and 48). Places close to the Equator (the tropics) will have high temperatures and a limited variation over the year, as the sun is always high in the daytime sky. At the other extreme, Polar Regions will always be cold. In between are the mid-latitudes or temperate regions, where temperatures will show a wide variation across the year (winter/summer). The climates are often described as Arctic, boreal, temperate, subtropical or tropical.

2. Rainfall (humidity)

Climates vary from wet or humid (with over 1500 mm a year), through semi-humid, to semi-arid and arid (less than 250 mm a year). Seasonal variation is also important. Rainfall may be distributed evenly throughout the year (UK) or distinct wet and dry seasons may be found. Globally, most regions receive their rainfall in their summer months, whereas in regions with a Mediterranean climate most rainfall is during the winter months.

Figure 48: World maps of mean sea level temperature (°C) for January and July

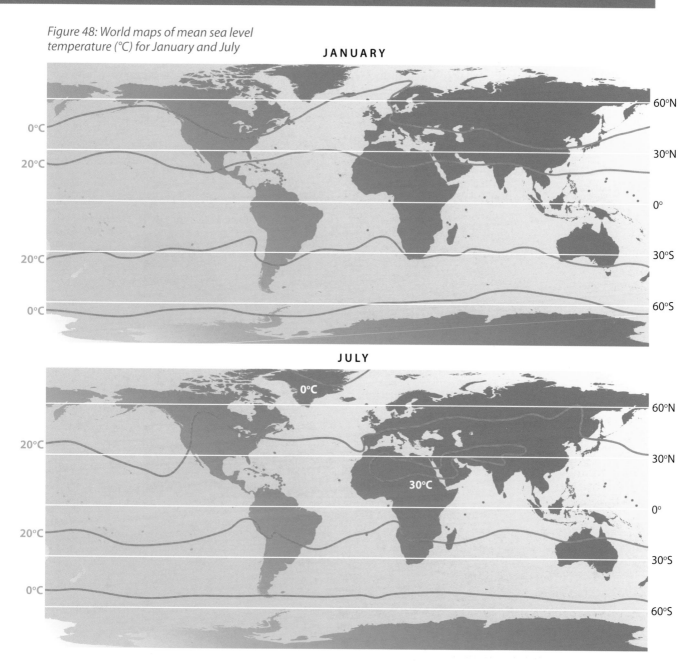

JANUARY

JULY

Figure 49: Global distribution of precipitation

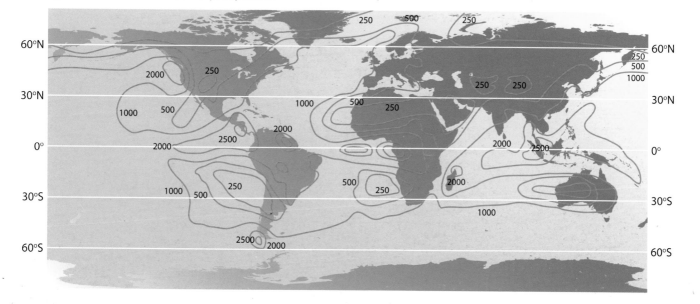

3. Height

Increasing height in mountainous areas causes a pattern of climate and therefore habitats similar to those found with increasing latitude. In short, temperature falls with height, hence Alpine environments are found even on the Equator on high mountains such as Mount Kilimanjaro.

Figure 50: The effect of altitude on climate and biomes in the Andes

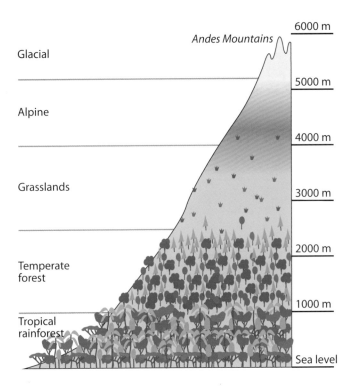

Figure 51: The global distribution of selected biomes

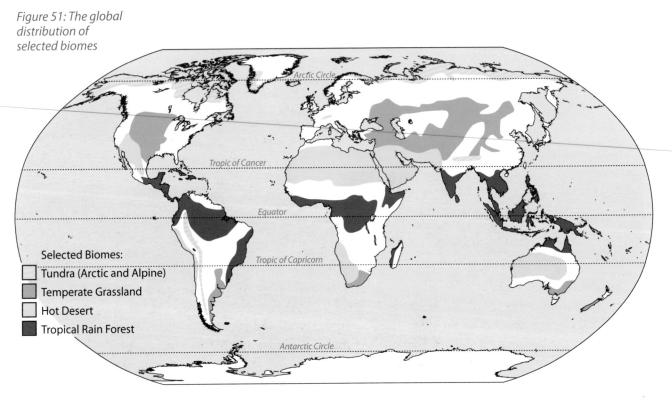

Selected Biomes:
- Tundra (Arctic and Alpine)
- Temperate Grassland
- Hot Desert
- Tropical Rain Forest

Mollisol/chernozem – the zonal soil associated with temperate grasslands

The zonal soil of the temperate grassland biome is called a mollisol or chernozem, meaning 'black earth'. The grass vegetation decomposes to produce neutral or slightly alkaline soils with a rich mull humus. This is quickly mixed into the upper horizons of the soil by earthworms and other soil fauna, giving a deep, dark, nutrient-rich A horizon. This fact, along with the soil's crumb or granular structure, makes the mollisol an extremely fertile soil type.

The topsoil is a dense mat of tangled roots, rhizomes, bulbs and rootstock, also known as the sod layer. Each winter the plants die back but regenerate from the underground bulbs and root systems. Roots of prairie plants can be longer than the plant is tall; Big Bluestem roots can measure 2 m and Switchgrass roots up to 3 m. It is estimated that two-thirds of prairie plant biomass is below the surface. Grass blades and some roots die in the autumn of each year and decompose, adding lots of organic matter to the soil. That is why the mollisol soil of prairies is so naturally nutrient-rich, fertile and productive.

In early spring the snowfall that has accumulated over the long winter months melts, causing mild downward leaching through the soil. However, later in the summer low rainfall and high temperatures draw water upwards in the soil by capillary action. These seasonal vertical movements of water help create and mix the deep and dominant A horizon top soil. Capillary action also deposits small nodules of calcium carbonate in the B (sub-soil) and C (parent rock) horizons. Commonly, the parent material is alkaline in nature, containing rocks such as limestone or a wind deposited material called loess, formed at the end of the last glaciation around 12,000 years ago.

Under natural conditions soil erosion, either by the intensive rainfall of summer thunderstorms or strong winds, is rare, as the deep rhizome and tap roots of the annual grasses are very effective at binding the upper soil horizon into a deep sod (Figure 60).

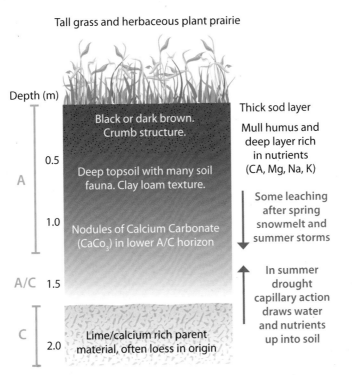

Tall grass and herbaceous plant prairie

Depth (m)

Black or dark brown. Crumb structure.

0.5

A

Deep topsoil with many soil fauna. Clay loam texture.

1.0

Nodules of Calcium Carbonate ($CaCo_3$) in lower A/C horizon

A/C   1.5

C

2.0   Lime/calcium rich parent material, often loess in origin

Thick sod layer

Mull humus and deep layer rich in nutrients (CA, Mg, Na, K)

Some leaching after spring snowmelt and summer storms

In summer drought capillary action draws water and nutrients up into soil

Balance of water annually: evaporation = precipitation

Figure 60: Zonal soil profile of a typical grassland mollisol/chernozem

A summary of the characteristics of the mollisols/chernozem soil of the temperate grasslands

The soil profile diagram (Figure 60) illustrates the characteristics of these soils that cover 7% of the earth's land surface and underlie the temperate grassland biome. Their nature reflects the regional climate and the cover of grass vegetation. The zonal soil profile is a very distinctive one:

Soil characteristics	Mollisol characteristics
Soil profile – Soils are described as having A (top soil), B (sub-soil) and C (parent material) horizons. The A horizon contains the bulk of a soil's nutrients that promote growth.	Mollisols have a deep topsoil (A horizon) with virtually no sub-soil (B horizon), as the organic material is mixed and there is slight leaching (washing down) and seasonal capillary action (moving up) of nutrients. This is helped by the active mixing by the soil fauna of worms, springtails, centipedes, mites etc (Figure 55).
Soil depth – reflects: a) the rate of decay and decomposition of organic material; b) the weathering rate of the parent rock; and c) time.	Mollisols are quite deep at 1–2m. Even though they are young, (post-glacial) soils, rapid weathering of the parent material and the annual addition of organic grasses (which die back each autumn with the onset of the cold, dry winter season) have quickly produced a deep soil.
Soil colour – reflects on the chemistry of the soil materials.	Mollisols are dark brown even black in colour, as they have high humus content (decayed organic matter). The annual grasses and limited leaching allow the rich organic material to build up within the soil. In the lower part of the A horizon, upward water movement (capillary action) commonly deposits white spots of calcium carbonate across the soil.
Soil acidity – measures the hydrogen ions in the soil. As these can replace the ions of nutrients, a high concentration (low pH 4) means growing conditions would be poor.	Mollisols have a neutral pH around 6/7 due to the limited spring leaching and the lime-rich parent material (commonly a wind-blown fine glacial material known as loess).
Soil texture – measures the mix of clay, silt and sand-sized inorganic particles in the soil. A loam texture is the best.	Mollisols have a loam or clay-loam texture, which allows them to drain well but also to retain essential nutrients and water for plant growth. The common parent material is a wind-blown deposit called loess which promotes this mixed texture.
Soil structure – describes how soil particles (humus and inorganic particles) clump together. The best soils have a crumb (bread crumb) structure.	Mollisols usually display the ideal crumb structure due to the abundant mull humus well mixed by water processes and the highly active soil organisms. This structure keeps the soil stable in the face of strong winds and heavy summer rain, allows the deep plant roots to grow freely, and retains water and heat in the soil.

The tundra biome

Figure 61: Permafrost – erosion on the Alaskan coast reveals the permanently frozen nature of the ground under tundra.

Source: U.S. Geological Survey

The climate and soil associated with the tundra biome

The tundra is one of the least productive biomes in the world due to its extreme climate. It lies between the taiga coniferous forest and the polar ice cap. Its name comes from the Finnish term meaning 'barren' or 'treeless plain', reflecting the harsh environmental conditions that prevent trees from becoming established. Where sub-soil temperatures remain below 0°C the ground becomes permanently frozen, this is termed permafrost. Permafrost is a deep layer of frozen ground, soil and dead plants. Between 300 and 1,600 m deep, it underlies almost one quarter of the earth's land surface. As permafrost locks moisture away, few plants can gain access to the water they need to thrive. There are only 50–60 days in summer when the upper soil

Figure 62: Caribou grazing on Alaskan tundra vegetation

Source: U.S. Geological Survey/photo by David Gustine

layer thaws out to form a temporary active layer. The species that can survive in this biome grow in large numbers across the ground surface, mostly with a low-lying cushion form. These include sedges, liverworts, berry-bearing plants (such as bilberry and crowberry), perennials with underground storage parts and small, thick evergreen leaves. Mosses and lichens are abundant, as their tissues cannot freeze and some can continue to photosynthesise at temperatures as low as –20°C. The ground-hugging pattern of most plants keeps them out of the fierce, chilling wind and when covered by snow, it acts as an insulator creating a warmer microclimate.

Only 1,700 species of plant are found in this biome, a low figure by comparison: even the UK has 3,000 plant species, not including fungi and lichens. This narrow range of vegetation supports a limited diversity of animals. Only 48 species of land mammal are found, including the large reindeer herds of Northern Eurasia (caribou in North America) which feed on lichens and reindeer moss. Other herbivores include

Arctic hares, picas, lemmings, musk-oxen and deer. Further up the food chain, the predators include wolves, wolverines, Arctic foxes, hawks and polar bears. About half of the animal species migrate south overland in winter, while those that remain hibernate for up to 10 months of the year.

The region supports few insect species but the vast numbers of black and deer flies, mosquitoes and biting midges make for miserable conditions in the tundra summer. Mosquitoes use a chemical called glycerol as antifreeze, allowing them to survive the winter under the snow cover. The marshy summer tundra provides a haven for migrating ducks, sandpipers and plovers attracted by the abundant food supply provided by swarming insects.

The climate of the tundra biome

This biome's climate consists of two distinctive seasons:

1. **Winter** – is bitterly cold and dark, with between 6–10 months below freezing (0°C) and average temperatures at –25°C. The sun barely rises about the horizon and plant activity is impossible.

2. **Summer** – during this season the sun shines all day and night in the 'Land of the midnight sun'. Temperatures reach 4–15°C during this 50–60 day long growing season.

    Annual precipitation totals are often lower than deserts, with between 130–250 mm but as evaporation and transpiration are low, there is water available in summer to supply plants. The region is swept by strong winds averaging 45–90 km/h and the lack of trees means there is little protection for the vegetation.

Spring and autumn are no more than short periods of transition between the two seasons (Figure 64).

The upper layer of the frozen soil thaws in summer but as water cannot percolate deeper due to the underlying permafrost, the summer landscape is dominated by lakes and marshes. Plant life will vary between these wet environments (rushes and sedges) and the drier, better drained slopes of hills (dwarf willow and grasses). Plants are highly adapted to the environment. They spread low across the ground surface in thick clumps and cycle through growth, flowering and reproduction very rapidly.

Gelisols – the soil classification of the tundra biome

The tundra is the world's youngest biome, having only been freed of a permanent ice cover 10,000 years ago when the last glaciation ended. This, along with the present harsh climatic conditions that slow down soil forming processes, means the soil is immature, shallow and infertile, with poorly developed horizons.

The soil profile diagram (Figure 65) illustrates a tundra soil 'at its best'. Soils are thin, with a lack of humus, as the organic material decomposes very slowly. Due to the frozen conditions of winter and waterlogging in the summer, soil organisms struggle to break down dead organic material to provide nutrients for new growth. Much of the soil is composed of angular rock fragments and little stratification can be identified. The upper soil only thaws for a few weeks in summer and during this time some decomposition may occur, producing a thin black mor humus. Acid in nature, this in itself does not help encourage soil organisms in the decomposition process. The waterlogged conditions that persist where soils are in low-lying hollows or at the base of hill slopes means chemical processes create a gley soil of blue-grey clay in the 'sub-soil' (B horizon) (Figure 65).

Figure 64: Climate graph for tundra biome, Point Barrow, Alaska, USA

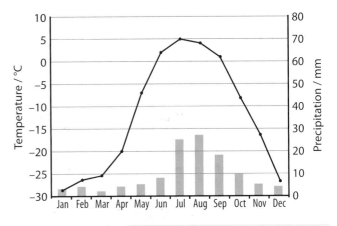

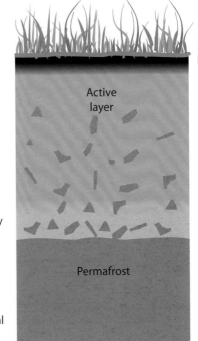

Figure 65: Soil profile of a tundra (gelisol) soil

Vegetation cover of grass and flowering plants

Black layer of mor (acid) humus

Angular fragments of rock

Blue-grey/grey clay or gley

Parent material

Mat of moss and lichen on surface

Active layer

Cold, acidic conditions mean organic litter decays slowly

Waterlogged during summer

Permafrost

Permanently frozen ground

Development and change in the Arctic tundra biome

Despite surviving in the harshest of environments, tundra vegetation is extremely vulnerable. A single tyre track left across a layer of tundra lichen and moss requires decades for recovery due to slow growth rates. Food webs are delicate and with limited variation in food sources, even impacts on one key species may be disastrous (Figure 66).

Figure 66: An example of a tundra food web

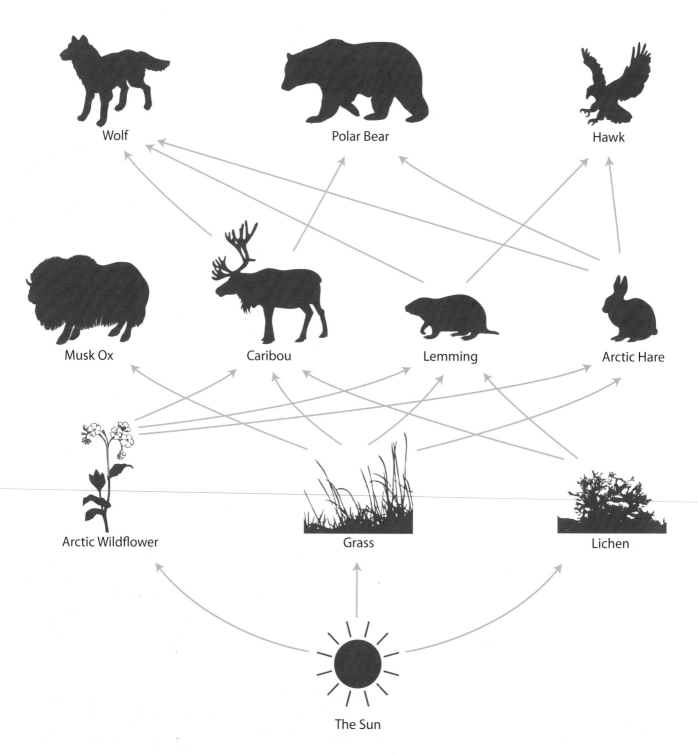

CASE STUDY: Alaska North Slope, USA, the actual and potential impacts of climate change on a tundra ecosystem

Background

Alaska North Slope, USA is the state of Alaska north of 68°N. It includes the National Petroleum Reserve Alaska (NPRA), the oil field developments at Prudhoe Bay and the Arctic National Wildlife Refuge (ANWR).

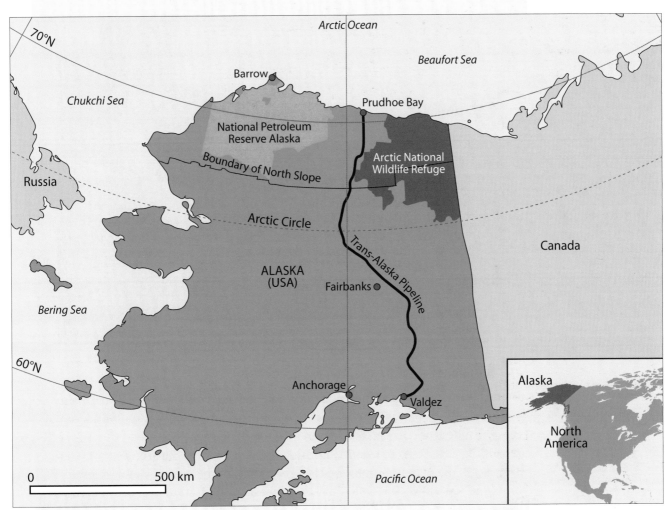

Figure 73: Alaska regional map

Prudhoe Bay and the oil industry

The impact of global warming in North Alaska is both significant and varied. It includes melting permafrost, retreating and thinning glaciers, eroding coasts and reduced sea ice cover, creating problems for species, including caribou, sea otters, salmon and polar bears.

Fifty years ago the discovery of vast oil reserves off the north coast of Alaska, near Prudhoe Bay, initiated an intense debate over the technological ability and economic significance of extracting the fossil fuel from this remote region. It was argued by many that the potential negative environmental impacts should ensure that no such development took place.

After five years of design and three years of construction, the $8 bn Trans-Alaska pipeline started shipping oil over 3,000 km from the north coast to the port of Valdez, where it is transferred to oil tankers. Around 1.4 million barrels of oil are moved daily along a pipe mainly elevated above the ground surface to avoid ice heaving and, as it is kept at 65°F, to prevent melting of the permafrost. The argument is that with good environmental planning, safe development on a large-scale is possible in the region. However, some would point to a calamitous oil spill from a tanker called the *Exxon Valdez*, on 24 March 1989, which destroyed ecosystems along Alaska's southern coastline, as evidence that such development is not compatible with this delicate environment and ecosystem.

Figure 74: View of the Trans-Alaska oil pipeline

Source: U.S. Geological Survey

The National Petroleum Reserve Alaska (NPRA)

In the NPRA the US government drilled and then abandoned 137 wells between 1944 and 1981 as part of an exploratory oil and gas program. By 2013 only seven of these abandoned wells had been properly plugged and reclaimed, and the remaining wells were an eyesore harming the environment. The reserve contains Teshekpuk Lake, an important nesting ground for numerous species of migratory birds and waterfowl. The NPRA has the highest concentration of grizzly bears and wolves in Alaska's Arctic. These prey on the abundant caribou that migrate in herds of over 500,000 animals. The reserve is home to Inuit settlements on its boundary, including Barrow, a town of 4,500 people that has become a focal point of climate change research.

In 1973, Barrow was selected as one of five key locations to form an atmosphere baseline for the global study of climate change. Since then, the following climate and environmental changes have been recorded:

1. Barrow's average temperature has risen by 3.5°C since 1973, higher than Alaska's 2°C increase.
2. The quantity of carbon dioxide and methane in the air has increased by 16% and 5% respectively in the last 40 years.
3. In spring, the winter snow now melts 10 days earlier than 30 years ago.
4. Winter sea ice is both thinner today and arrives later in the year.

5. Reduced sea ice has caused coastal erosion rates to increase and consequently several villages have been moved further inland for safety.

6. Species new to the area are regularly seen, including insects (spruce beetle) and bird life.

7. Summer lakes have vanished, draining away as the permafrost melts and leaving migrating birds without water and a breeding ground.

Figure 75: The small town of Barrow on the north coast of Alaska.

Source: Dave Cohoe

The Arctic National Wildlife Refuge (ANWR)

Alaska's Arctic National Wildlife Refuge is a 19.6 million acre wildlife sanctuary, located in the remote corner of the state. The landscape in the north is a wide expanse of tundra, with numerous marshes and lakes. This is a fragile permafrost environment, where human impacts on the landscape can take decades or centuries to disappear. The ANWR's Arctic Ocean coastline plain has been described as the largest intact, naturally functioning Arctic ecosystem. It contains:

- the most significant onshore area for polar bears in the USA.
- key nesting and breeding sites for 160 bird species.
- the breeding ground for a caribou herd of over 130,000 animals. These are the main source of clothing, food and medicine for the Gwich'in Indians, one of the world's few remaining subsistence cultures.
- Arctic foxes, wolves, wolverines, grizzly bears, whales and other key Arctic species.

Actual and potential change
Local climate

Average Alaskan temperatures have increased by 2°C in the past five decades, with winter increases being the most pronounced at 3.5°C. Alaska's North Slope has warmed dramatically, especially in late summer when ice cover is at its lowest and the summer heat absorbed by the ocean is emitted into the atmosphere from the larger expanse of open water. This warming has serious consequences: the presence of open water, rather than solid ice at sea or on rivers, disrupts routes traditionally used for travel; the land may consist of soft and failing soil, instead of hard-frozen ground; and precipitation falls more often as rain than snow.

Sea level

Melting sea ice raises many environmental issues but it does not raise sea-levels, as water displaces the same mass when frozen or liquid. An ice cube in a drink may dilute the drink when it melts but the water level remains unchanged. However, Alaska holds 11% of the world's mountain glaciers and currently contributes about a quarter of the world's mountain glacier meltwater. Annually Alaskan glaciers are losing 75 billion tonnes of ice, and many are expected to disappear entirely by the end of the century. This not only affects sea levels. This increasing quantity of fresh water changes both the sea's salinity and the flow of currents circulating in the Arctic Ocean. A US Geological Survey report revealed that Alaska's North Slope has lost an average 1.5 m of coastline to erosion annually since 1950. This places both village settlements and the oil field infrastructure at risk. Twenty-six communities suffer erosion problems that need an immediate response: some require hard engineering protection, others total relocation.

Figure 76: Erosion rates are increasing along Alaska's coastline

Source: U.S. Geological Survey/photo by Benjamin Jones

Water quality

Thawing permafrost and eroding soils release stored mercury, which adds to the pollution carried by the air and oceans from distant industrial and coal burning operations. Chemical pollutants, such as pesticides held in sea ice, are being released into the water as the ice thaws. This process helps to explain the increasing acidity of Arctic Ocean regions, including the Bering Strait. Recent changes in the Beaufort Sea mean the water has too little calcium to support shell-building organisms. The waters of the Chukchi and Bering Seas are expected to become similarly acidic in the near future.

Wildlife

The polar bear has become the icon of global warming and was the first animal to be given Endangered Species Act protection because of climate change. But biologists claim that many other species are at risk as sea ice disappears. Pacific walruses are also candidates for Endangered Species Act protections, as in recent years they have been forced to crowd onto the Chukchi Sea shoreline because they can't find floating sea ice in late summer. Likewise, bearded and ringed seals are listed as threatened. Land animals too have problems with ecosystem change. The northward expansion of shrubs is bad for caribou, musk oxen and other animals that depend on tundra habitat.

As a result of these environmental factors, the diversity of plants and animals in Breen Wood is quite limited and the trees tend to grow slowly and have a stunted appearance. Two-hundred-year-old oak trees on the higher slopes of Breen Wood have reached only half the height they would have attained in lower lying areas.

The biotic environment
Autotrophs

As mentioned, the dominant tree species in Breen Wood's vegetation cover are the oaks but also common is another deciduous tree, the downy birch. Below the canopy is a shrub layer which is mainly composed of smaller tree species, such as rowan, hazel and holly. Beneath and around these is the field layer of bilberry, bramble and numerous species of fern. On the soil surface, great wood-rush, wood-sorrel, mosses and ferns form a blanket cover.

The appearance of the deciduous woodland changes greatly with the seasons. In early spring the bare trees allow sunlight to reach the forest floor and for a few weeks bluebells and anemones produce a sea of blue and white flowers, whereas in summer the complete leaf canopy restricts the ground plant cover to grasses and ferns (Figure 85). Fungi are common, such as the Birch Bracket (Figure 86), while mosses, lichens and ferns often cover the branches and trunks of dead and living trees.

Figure 85: Spring and late summer in Breen Wood

Figure 86: Birch Bracket fungi in Breen Wood

Heterotrophs

Once more, the harsh environment is a limiting factor on the animal life of these woods. Many insect species are found in and around the vegetation, including at least 14 species of caterpillars, later to be butterflies, the Orange Tip and Speckled Wood. The native red squirrel is the herbivore most closely associated with the oak woods, although the introduced larger grey squirrel species threatens its future.

Other heterotrophs are the woodland birds, including the Goldcrest, and Great, Blue and Coal tits that forage the forest for their diet of seeds and insects. Goldcrests are the smallest bird and feed on insects in the tree canopy. All these small birds can become the prey of larger carnivores passing through, such as buzzards and sparrow hawks. On the ground, the largest mammals are the badger, the stoat and the fox. All three are

mainly nocturnal animals. However, stoats are carnivores (eating rodents, birds and eggs), while foxes and badgers are omnivores. Foxes prey on birds and small mammals, such as squirrels, but also eat grass and fruit. Badgers consume earthworms, small birds, fruit, nuts and the root bulbs of plants.

Nutrient cycling

The harsh abiotic habitat of cool temperatures and acidic soils makes survival difficult for many species of soil organism. Few earthworms are found and consequently litter, such as the autumn leaf fall, may take years to fully decompose and return their nutrients to the soil. The nutrient input from the underlying rock is low, leaving the rainfall as the main source of new chemical nutrients. As noted, the release of nutrients from the litter to soil stores is slow but with autumn leaf fall a large flow from the biomass to the litter store does take place. Leaching of nutrients down through the soil beneath Breen Wood carries nutrients beyond the reach of the plant roots and away in the small streams that drain the area.

Figure 87: Fungi decomposers on a fallen oak branch

Figure 88: Small nesting boxes for the spring season at Breen Wood

Figure 89: Layer structure of vegetation in the oak woodland, Breen Wood

Vegetation succession

As any gardener will know, when presented with a newly-available area of rock or soil, plants will quickly invade. This process is called **colonisation** and the first plants to arrive are termed **pioneers**. This is the first step in a sequence of changes that will eventually produce a balanced community of plants called the **climatic climax vegetation**. This is the process of **succession**; the whole sequence is called a **sere** and each step in the process is a **seral stage**.

As the name suggests, a climatic climax community is the group of plant species that are best adjusted to the climatic factors that exist. As noted earlier, most of Ireland would be forested if left to nature, so even an abandoned garden would undergo succession to reach deciduous woodland in a century or so.

The succession process helps explain why weeds can invade a newly-available site but shrubs or trees cannot be established until later in the sequence.

The plants that most easily adapt to the unique environmental conditions of the site will colonise a location. Each plant species in a community has a specific range of tolerance of the abiotic conditions. Climate and soil factors are the most important influence over the successful establishment of plant communities. These include sunlight, moisture, soil fertility and stability:

1. Not only is the *amount* of sunlight available important but also its *duration* and *quality*. The duration of sunlight affects the flowering of plants. The intensity of light affects photosynthesis and the growth rate.
2. The *availability of water* is important for the survival of most life forms but plants require water for a number of life processes – germination, growth and reproduction.
3. The presence and depth of the soil, along with the range of nutrients available, will influence which plants thrive.

In their turn, the pioneer community of plants and each successive seral stage alters the environment in such a way to permit new species to occupy the environment. These alterations of the ecosystem's abiotic environment involve small but progressive changes in the microclimate and soil conditions of the site. Plants provide shade and protection for others, and after they die their decomposition improves the soil's structure and its potential supply of nutrients.

An example: Following a landslide, rock surfaces are exposed for the first time to the atmosphere. On this bare inorganic surface there is no soil, so most plants cannot grow. However, pioneering lichens and simple mosses may survive by obtaining nutrients from the rock itself and rainwater. These plants help to weather and break-down the rock surface, allowing other species to be established. Plant roots help to bind the loose rock fragments as a thin soil, and plant stems and leaves shade the surface from the bright sun and shelter it from the wind. When these plants die they add organic material that holds nutrients and water in the soil. Over time more species are added, but the pioneer species of lichens and mosses die out as the taller plants shade out their access to sunlight. And so succession continues.

A climatic climax community is the result of a long period of plant succession. Such communities usually exhibit a good deal of species diversity and thus are relatively stable systems. If a succession continues uninterrupted by natural events or human activity from pioneer to climax, it is termed **primary succession**. On the other hand, if it is disturbed this renews a succession sequence as **secondary succession** (Figure 99). Natural disturbance may involve fires, landslides, flows of lava or volcanic ash fall, severe floods or storms. In 1987 a single storm brought down over one million trees in South East England. Human disturbance related to tropical deforestation has caused secondary succession of plant communities in the tropical rainforest.

Figure 90: Accidental and managed interruptions to succession

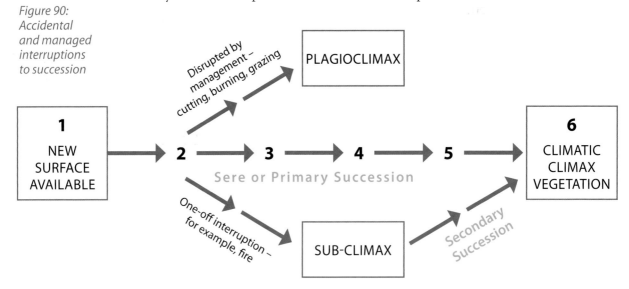

Plagioclimax

Plagioclimax is a term used to describe the vegetation community that develops when a climatic climax or succession is changed by human activity. The term implies that it is not just a disruption to the natural vegetation but rather a management system that prevents any natural change and 'arrests' the vegetation system.

Most landscapes in the British Isles today are plagioclimaxes, including school playing pitches, farm fields and planted forests. Clearance by cutting, burning and grazing are the three most common human activities which create plagioclimax communities.

Figure 91: Muirburn on the Fleak, Askrigg Common, North Yorkshire

Much of upland Britain is covered by moorland, often dominated by a few species, such as heathers and rough grasses. In nature most of these areas would undergo succession to woodland, but the use of the land for animal grazing (sheep and cattle) and burning to improve pasture, prevents the growth of tree saplings. In some upland areas the carefully managed burning of heather moorland (the muirburn) is designed to maintain the population of red grouse. The moor landscape then appears as a patchwork of different colours, with heather at different ages. In the shooting season, game bird hunters pay large sums to shoot grouse over these highly managed moorlands. Figure 91 shows the patchwork pattern of vegetation produced by years of Muirburn management. In the foreground is a shooting butt, one of a line, stretching to the horizon.

There are four main starting environments for primary succession: two land based and two water based. It is suggested that under natural succession most of the island of Ireland would reach a climatic climax vegetation community of deciduous, mainly oak woodland.

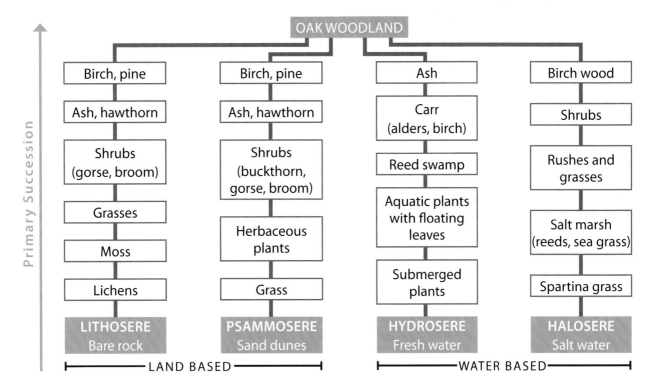

Figure 92: Primary successions in a West European climate

Two sample succession case studies

The north of Ireland provides a remarkably broad range of environments in which succession processes may be studied. Some are particularly useful in that many seral stages can be studied at the same location. In places along the coast, a sequence of sand dunes develop behind wide sandy beaches, and elsewhere, freshwater lakes and ponds are gradually being filled in by sediment and plant debris.

Note:

The specification requires the study of only **one** succession case study.

Each of these dynamic landscapes provides an opportunity to witness the colonisation progress from the invasion of the pioneer plants through various seral stages to the climatic climax vegetation.

CASE STUDY: Hydrosere succession in Hollymount National Nature Reserve, County Down

Around the edges of fresh water ponds or lakes, seral stages of succession can be seen. Such water bodies are gradually being filled in by sediment, washed from surrounding slopes or brought in by streams, making the shores shallower and more fertile. In the deeper water, submerged and floating plants, duckweed and water fern are found, along with plants rooted in lake bed sediments, such as water-lilies. In the shallows, plants such as rushes, reeds and sedges dominate. These are hydrophytes – plants that can absorb oxygen through their roots even in waterlogged soils. Such marsh plants grow out into the lake, their roots trapping silt, and when they die, their remains gradually build up on the floor of the lake. When the sediment rises to the water level, the environment is described as a fen or

Figure 93: Pond succession, County Down

carr, and water-tolerant tree species such as alder and willow take root. The succession moves from the aquatic (water) based section of its sequence to a terrestrial (land) based one. As the lake shrinks, the drier soil conditions allow other trees to become established, including ash. Eventually, through a sequence of seral stages, a mixed oak woodland climax vegetation cover can be attained.

Figure 94: Model of hydrosere succession

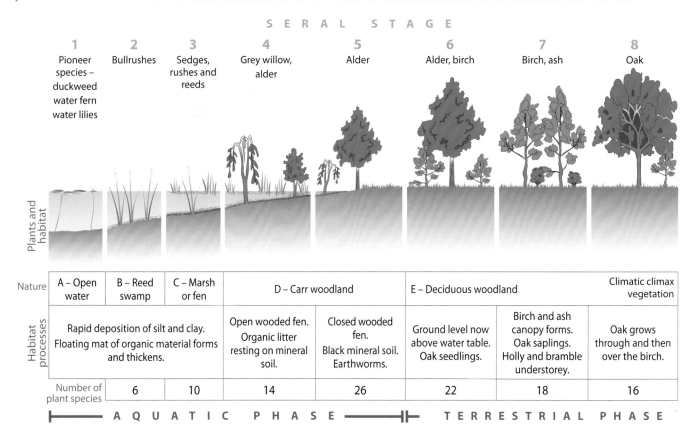

SERAL STAGE

1	2	3	4	5	6	7	8
Pioneer species – duckweed water fern water lilies	Bullrushes	Sedges, rushes and reeds	Grey willow, alder	Alder	Alder, birch	Birch, ash	Oak

Plants and habitat

Nature	A – Open water	B – Reed swamp	C – Marsh or fen	D – Carr woodland		E – Deciduous woodland		Climatic climax vegetation
Habitat processes	Rapid deposition of silt and clay. Floating mat of organic material forms and thickens.			Open wooded fen. Organic litter resting on mineral soil.	Closed wooded fen. Black mineral soil. Earthworms.	Ground level now above water table. Oak seedlings.	Birch and ash canopy forms. Oak saplings. Holly and bramble understorey.	Oak grows through and then over the birch.
Number of plant species	6	10	14	26	22	18	16

├────── A Q U A T I C   P H A S E ──────┤├ T E R R E S T R I A L   P H A S E

Figure 95: Hollymount location sketch map

Around Hollymount NNR (National Nature Reserve), a few kilometres west of Downpatrick, a hydrosere succession can be observed. Here lies one of Ireland's best examples of an alder carr woodland environment, with its damp vegetation community. In common with much of County Down, the area is covered by low, elongated, oval shaped hills of glacial origin named drumlins. These produce a chaotic drainage pattern, including inter-drumlin hollows in which rainwater gathers. In the Hollymount area there are inter-drumlin lakes, partially filled-in ponds, marshes or fens, and dry drumlin slopes; in fact, all the seral stages associated with a fresh water hydrosere occur in this location.

Near the lake shore water-lilies and duckweed, with their large and tiny leaves, respectively, are found merging with the marsh plants – soft rush, great reedmace and great pond sedge. In the wet woodlands the ground flora includes reed canary-grass and in particular tussock sedge. Above these is a dense canopy formed by alder and grey willow trees.

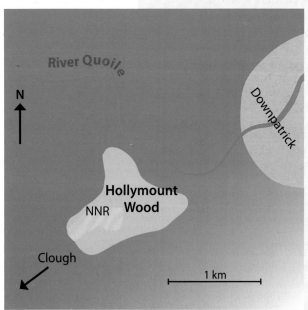

River Quoile

N

Downpatrick

Hollymount Wood

NNR

Clough

1 km

Location	Latitude	Temperature low and high	Annual Range
Oslo	60°N	–5° to 16°C	21°C
Rome	42°N	8° to 24°C	16°C
Libreville	0°	24° to 27°C	3°C

Figure 108: The impact of seasonality on annual temperature range

Other factors that locally influence temperature include:

- **Prevailing winds and ocean currents** – will cause local heating or cooling. Northerly winds bring colder air to Ireland and the UK, while the warm ocean current known as the North Atlantic Drift brings warming tropical water to the same islands.

- **Aspect** – is the angle of slope direction in relation to the sun. In Europe south facing slopes will generally be warmer as they face the sun.

The impact of many of these factors can be seen on the world temperature maps (Figure 109) but one in particular cannot. Which of these factors has been deliberately omitted from the maps below?

Figure 109: World maps of mean sea level temperature (°C) for January and July

JANUARY

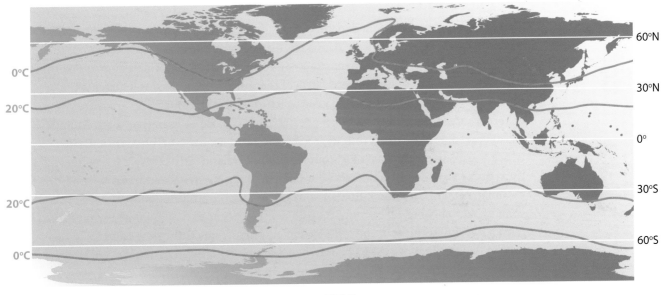

JULY

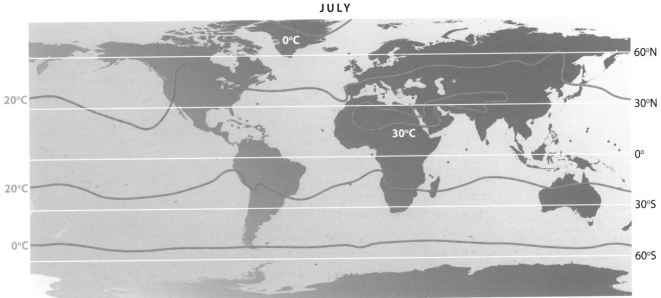

Vertical heat transfer

Heat energy is transferred vertically from the earth's surface into the atmosphere, cooling the earth while warming the air. There are several mechanisms for this transfer, including:

- radiation – the earth's surface emits long wave energy which radiates out into space.
- conduction – energy transferred by contact.
- convection – rising currents of warm air into the atmosphere.

Horizontal heat transfer

The two natural routes for exchanging heat energy across lines of latitude to remove the imbalance are:

- winds in the atmosphere.
- currents in the oceans.

Research suggests that winds are the more important mechanism but that ocean currents still account for up to 30% of the energy exchange.

1. Ocean currents

These are complex flows that depend on heat energy and variations in salinity but the most consistent ocean currents form a circular pattern in each of the world's oceans. Ocean currents that flow away (north or south) from the Equatorial region are warm and these are found in the western parts of the world's ocean basins. Currents that move towards the Equator carry cooler water towards the tropics and are therefore cold currents, usually in the eastern ocean basins (Figure 110). In the North Atlantic a warm current crosses over from west to east to produce the **North Atlantic Drift**, which warms the west coast of Europe, including Ireland (50°–55°N). Meanwhile, off the coast of Canada, the cold Labrador current brings icebergs and freezing conditions to Canada at latitudes much further south. The *Titanic*, built in Belfast at 54°N, was sunk by an iceberg at 42°N, more than 14 degrees closer to the Equator! Perhaps the most powerful example of the influence of ocean currents is the flow of water in the central Pacific, known as the El Niño. Across the world this occasional event triggers unusual and extreme weather, including storms and droughts, often with large-scale impacts on society and the economy (Figure 129, page 119).

Figure 110: The main ocean currents

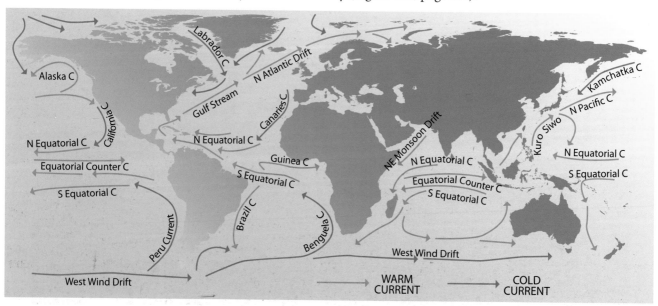

Low pressure systems – the mid-latitude depression

The depression, or low pressure system, is a huge mass of spiralling air up to 2,500 km across, which involves two contrasting air masses. More than any other feature, the depression illustrates the exchange of energy by warm, moist air from the tropics with the cold air from the Polar Regions. The UK and Ireland lie in a zone that is dominated by the west to east passage of about 30–35 depressions each year. Recently weather forecasters have started to name these, during the winter, in alphabetical order. The first so named was Abigail (as in 'a big gale'), in November 2015.

These powerful features form over the Northern Atlantic and pass, often within 24 hours, over the UK and Ireland. Each depression consists of a body or sector of warm air (**Tm air**), from the tropical region of the Atlantic, being gradually surrounded and lifted off the ground by cold polar air (**Pm**). The whole system has surface winds spiralling anti-clockwise, inwards towards the centre.

The contact zone between the warm and cold air is called a **front**. The front is described as being **warm** if, when it passes, you enter warmer air. It is described as **cold** if, when it passes, you enter cold air. At a front the cold, polar air undercuts the warm air, forcing it to rise. As the air rises it expands, cools, condenses and cloud forms, eventually bringing rain (cyclonic rainfall). Fronts are not vertical but slope upwards; the warm front in advance of the depression, and the cold front trailing behind the system.

Origins

Depressions and their weather patterns are formed by the fast-moving and wandering upper troposphere winds called the **upper westerlies** (also known as Rossby waves). These westerly winds circle the earth above the mid-latitudes and are the result of the large temperature difference between the tropical air to the south and cold polar air to the north. The fastest flow within these westerly winds is the Polar Front Jet Stream, which reaches speeds of over 200 km/h in summer and 450 km/h in winter. It appears that where these winds slow down (decelerate), air is forced downwards towards the earth's surface to form high pressure anticyclonic systems. Where they speed up (accelerate) they draw surface air upwards, creating low pressure systems at ground level (Figures 122 and 123).

The metcheck website provides an interactive map with useful information on the role of the upper atmospheric air flows: http://www.metcheck.com/UK/jetstream.asp

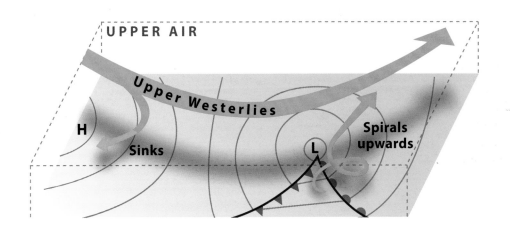

Figure 122: The relationship between the upper westerly winds and surface weather conditions

KEY:

Wind
(speed in knots)

◎ Calm

⌐○ 1–2

⊢○ 3–7

⊨○ 8–12

⊫○ 13–17

For each additional
half-feather add 5 knots.

⟋○ 48–52

Cloud
(amount in oktas)

○ 0

◐ 1 or less

◑ 2

◕ 3

◑ 4

◕ 5

● 6

◑ 7 or more

● 8

⊗ sky obscured
(usually fog)

⊠ missing or
doubtful data

Weather

= mist

≡ fog

• drizzle

⦂ rain and drizzle

• rain

⁑ sleet

✳ snow

⁎ rain shower

⁑ rain and snow shower

▽ snow shower

△ hail shower

ϟ thunderstorm

Temperature (in celsius (°C))

*Denoted by a number
beside the weather symbol*

Pressure (in millibars)

*Presure is shown by isobars.
These are lines joining
places of equal pressure.*

Fronts

⬤⬤⬤— Warm front

⬤▲⬤▲ Occluded front

▲▲▲ Cold front

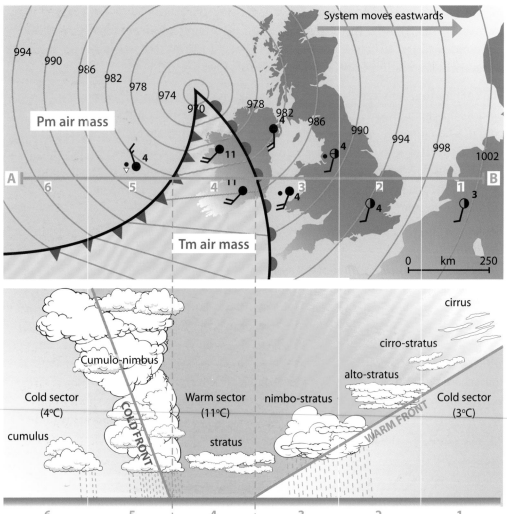

Figure 123: Weather map showing a mid-latitude frontal depression

	After cold front	As cold front passes	Warm sector	As warm front passes	As warm front approaches	Before warm front
	6	5	4	3	2	1
Cloud type	Fair weather cumulus	Towering cumulo-nimbus	Dull, low, flat stratus	Dense nimbo-stratus	Lower, thicker alto-stratus	High altitude cirrus and cirro-stratus
Rainfall	Clearing showers	Heavy down pour	Relatively dry	Persistent rain	Rain commences	Dry
Temperature	Cold	Falls rapidly	Warm	Warming	Cold	Cold
Wind	North or north west	More northerly	Moderate westerly	Strong more westerly	Increasing from south west	Light southerly
Air pressure	Rising	Rising	Steady	Falling	Falling 1002–988	Falling High (1002)

Synoptic charts

Synoptic charts show air pressure at the surface using isobars, lines of equal pressure. These are normally marked at 4 millibar (mb) intervals to clarify the patterns and pressure gradient across the area. In addition, the surface boundary line between air masses is marked as a front – cold, warm or occluded. Figure 112 (page 98) is an example of a synoptic chart.

The nature of a mid-latitude depression over the British Isles

Depressions are areas of relatively low pressure with central surface values below 1000 mb. They are marked by strong winds spiralling anticlockwise at the surface and rising up into the troposphere. Depressions are associated with bands of clouds and rainfall as a result of this rising air. On **synoptic charts** (weather maps) they are shown by a set of closed circular isobars (see Figure 112 on page 98 and Figure 123), which are tightly spaced, indicating a steep pressure gradient giving powerful winds. Depressions involve two different air masses, most often Polar maritime air moves around and lifts a sector of warmer Tropical maritime air. Depressions regularly pass across from west to east, arriving first on the wild Atlantic west coast of Ireland before crossing the islands on into the North Sea and continental Europe. Each separate front, warm and cold, is associated with rising air and consequently the development of both cloud and rainfall. In a later stage of a depression's life-cycle, the cold air lifts the warm air off the surface completely, giving a single occluded front. Fronts mark contrasts in air temperature and pressure, consequently they are often associated with strong, gusty winds that can disrupt transport systems and damage buildings and infrastructure.

Weather sequence

While no two depressions are exactly the same, many have similar life cycles and bring a common pattern of weather. Over Ireland and Great Britain most depressions travel from the west/south west and sweep across the region. Depressions bring cloud, rain and strong winds, especially when the warm and cold fronts pass over. Depressions last for 5–10 days, by which time the cold air lifts the warm tropical air off the ground (an **occlusion**) and the whole system dies away. Depressions rarely come singly; often several follow each other as a 'family' of depressions. Figure 123 shows the sequence of weather associated with depressions, including changes in cloud type, coverage, rainfall, and wind speed and direction. This sequence and the reasons for the changing conditions need to be known.

Figure 124: Contrasts in summer and winter weather under an anticyclone

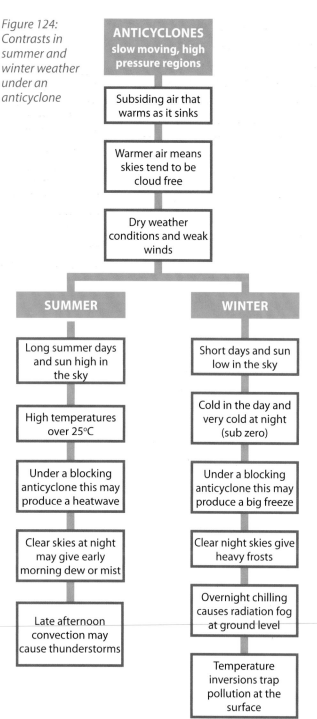

High pressure systems – anticyclones

Anticyclones are large, 3,000 km in diameter, masses of subsiding air sinking from a height of 8 km in the troposphere towards the surface. At ground level the air slowly spirals away in a clockwise direction (northern hemisphere). As the air descends it warms, and consequently the air is dry, so clouds and rain rarely form. Anticyclones are larger than depressions, involve only one air mass and move more slowly, lasting for several days or even weeks. The pattern of weather associated with depressions is similar throughout the year but anticyclones produce contrasting surface weather in summer and winter (Figure 124).

Formation

As mentioned earlier, anticyclones, like depressions, are the product of westerly winds moving in the upper troposphere. Where these air flows decelerate, air is forced to move downwards (subside) to form a surface high.

Anticyclonic weather

In nature summer and winter anticyclones:

- are marked by a central area of subsiding air that warms up, preventing rising air.

- produce calm or gentle surface wind that blows outwards and, in the northern hemisphere, clockwise.

- have clear skies allowing long hours of solar insolation.

- create dry conditions as uplift is restricted by the subsidence.

- may last for several days or even weeks, under a blocking system.

The impact of summer anticyclones

The weather produced under a summer anticyclone includes hot, sunny days, typically 25–30°C in southern parts of Ireland and Great Britain. At the same time, night time cooling of the ground chills the surface air, producing early morning dew or mist that normally evaporates as temperature rises. While calm conditions are normal under high pressure (isobars are widely spaced), local winds can develop including on-shore winds during the day. This is one of the reasons why people flock to the coast on very warm summer days, as on-shore, cool winds from the sea moderate the temperature locally. During long summer days, with up to 17 hours of daylight at this latitude, by late afternoon hot air at the surface may rise by convection, as a thermal, to form vertical cumulus or cumulo-nimbus clouds, causing thunderous rain or hailstorms. It is these intense storms that cause flash floods in August and disrupt afternoon play in summer sports, such as tennis or cricket. In a blocking anticyclone situation, a large area of high pressure remains in place for days or weeks, forcing the depressions that move in from the Atlantic away to the north or south of their usual path. An example was the 'late' or Indian summer of October 2015, when Northern Ireland enjoyed average temperatures 0.5°C above normal, 120% of normal sunshine and only 59% of average rainfall. Under these circumstances, and if the air in the anticyclone is Tc (Tropical continental) in origin, high temperatures and rapid evaporation can produce heat waves and drought. Both are at least inconvenient and potentially dangerous to health and the economy. Forest and heath fires can threaten ecosystems and property. The light breeze or calm conditions of blocking anticyclones and the trapping of cool air near the ground can trap pollution. Vehicle exhaust and industrial emissions may react chemically with sunlight to produce photochemical smogs. These, and the concentration of ozone at ground levels, can be a hazard to the health of young children and people who suffer from asthma or similar respiratory conditions. A health alert in southern England in April 2014 was linked to a concentration of Saharan dust and European Industrial pollution trapped in an anticyclonic system.

The following website links provide daily weather summaries for 1 October 2015 and 1 April 2014:

http://www. metoffice.gov.uk/ media/pdf/t/g/ DWS_2015_10.pdf

http://www. metoffice.gov.uk/ media/pdf/l/7/ DWS_2014_04.pdf

The impact of winter anticyclones (Figure 125)

The clear sky caused by warming subsiding air produces very different weather conditions in a winter anticyclone. Long nights of radiation heat loss often produce sub-zero temperatures, hoar frost and persistent fogs. These conditions can persist for days under a blocking system, as the sun is too weak to warm the land surface. The disruption to everyday life can be severe. In homes, water in unprotected pipes may freeze and, as the ice expands, the pipes break. Once this ice thaws, homes are flooded. Elderly people are at risk from hypothermia, as they may not be able to afford adequate

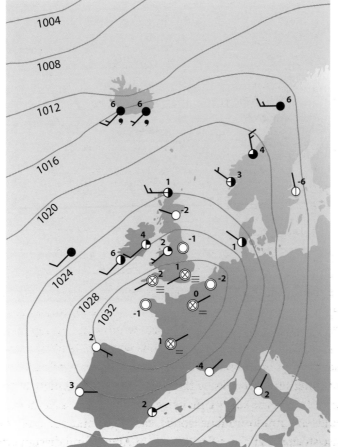

Figure 125: A winter anticyclone over western Europe

The following Met Office website link provides a detailed description for the analysis of weather maps or synoptic charts: http://www.metoffice.gov.uk/media/pdf/a/t/No._11_-_Weather_Charts.pdf

heating in their homes. Icy pavements force people to stay indoors or risk injury. Transport by road, rail and air may be curtailed by both ice and fog. The worst effects are normally seen in inland areas because at coasts the proximity of the sea helps to moderate the temperature. In both central Ireland and England several days of persistent fog and sub-zero temperatures are not uncommon between December and February. The early winter of November–December 2010 was the coldest ever recorded across much of the UK, with record low temperatures set in both Wales (–23.3°C) and Northern Ireland (–18.7°C). This was largely due to the position and shape of the upper westerlies that caused Arctic airs flows to dominate the weather during those two months. In contrast the winter of 2015–2016 was one of the mildest ever recorded, as a series of intense low pressure systems swept across the country giving very mild conditions at Christmas and the New Year. Unfortunately, these same storms brought exceptionally heavy rain, which caused widespread and repeated flooding in Cumbria, Wales and Yorkshire.

Interpreting weather systems using synoptic charts and satellite images

The combination of maps recording surface pressure and images taken from satellites orbiting the earth allow even non-experts to study the links between pressure systems and the weather.

Synoptic charts

These are maps designed to show surface weather features using isobars to indicate the variation in surface pressure conditions, weather fronts to locate the surface boundaries between air masses and weather stations showing a summary of current weather conditions.

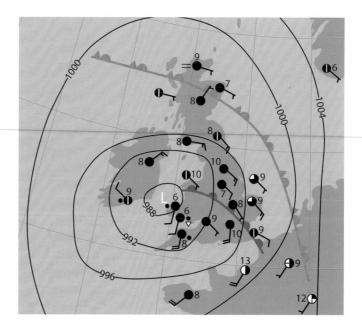

Figure 126: Sample synoptic chart

Satellite images

Geostationary satellites orbit at 22,300 miles above the earth's surface to provide global images of weather features. Near Polar orbiting satellites circle 500 miles above the earth, giving surface images every six hours. Satellites record images using visible light during the day and infrared images day and night. Infrared photographs allow temperatures of cloud and surfaces to be accurately shown. The lower orbiting satellites also use microwave detection through clouds to identify precipitation, temperatures at different levels in the atmosphere and surface characteristics.

Figures 127 and 128 are examples of the combined use of synoptic charts and satellite images in the study of surface weather patterns.

Winter storm – 9 December 2014

The synoptic chart in Figure 127 shows a deep Atlantic depression centred west of Iceland, with intensely packed isobars on its southern edge and an occluded front stretching east to the UK. The warm front of this depression extends from the north of Scotland along the west coast of Great Britain, while the cold front trails away to the west of Ireland. The warm sector lies between the two, covering Ireland and the Irish Sea. In the West Atlantic there is a small high pressure system and in the east high pressure, with widely spaced isobars, covers France.

Figure 127: Synoptic chart (below) and satellite image (right) showing a low pressure system 9 December 2014

Source: Synoptic chart data from the Met Office, Crown copyright 2016. Contains public sector information licensed under the Open Government Licence v1.0.
Satellite image copyright 2016 EUMETSAT, data from the Norwegian Meteorological Institute.

12:00 UTC 9 December 2014

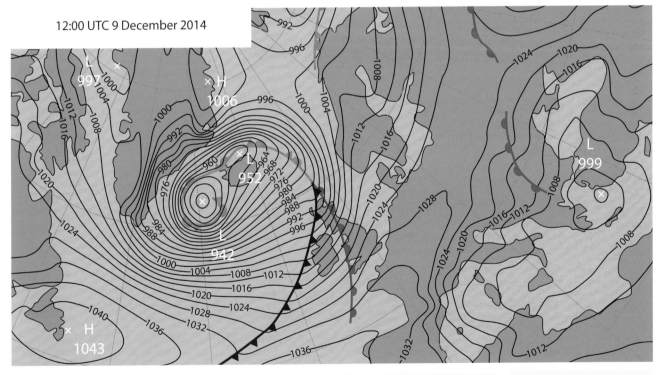

Exercise

1. Study Figure 127. Explain how the synoptic chart and its description above helps to explain the cloud pattern and weather conditions shown by the visible light satellite image.

The following website link shows footage of weather conditions on 9–10 December 2014:

https://www.youtube.com/watch?v=rmJxfP01O6Q&feature=youtu.be

Summer heatwave – July 2013

In July 2013 Ireland and the UK experienced a spell of hot, sunny and dry weather when an area of high pressure, known as a 'blocking anticyclone', became established across the islands. Many areas, particularly in the south and west, experienced almost unbroken sunshine. Daily temperatures exceeded 28°C widely every day from 13–19 July inclusive.

Figure 128: Synoptic chart (below) and satellite image (right) showing a high pressure system 19 July 2013

Source: Synoptic chart data from the Met Office, Crown copyright 2016. Contains public sector information licensed under the Open Government Licence v1.0.

Satellite image courtesy of NEODAAS/University of Dundee

The visible light satellite image for 19 July 2013 shows almost unbroken sunshine across the UK. Some low cloud and fog is seen affecting the North Sea coast and around northern Scotland.

12:00 UTC 19 July 2013

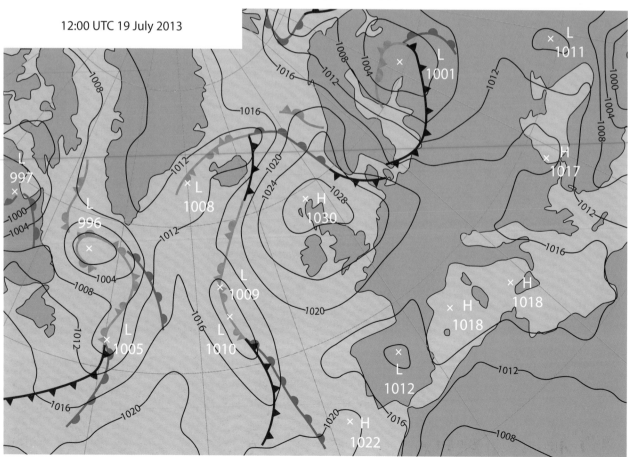

Synoptic situation at 12:00 UTC 19 July 2013, with high pressure established over the UK.

Exam Questions

1. Study Resource A, which shows changes in seven weather elements associated with the passage of a warm front in a mid-latitude depression.

   Describe and explain, with reference to **any four** of the elements, the weather associated with the passage of the warm front at 3pm. [8]

   Question from CCEA AS1 Physical Geography paper, January 2011, © CCEA 2016

Resource A
Adapted from Geography Review Vol 20, No 1, Sep 2016
For copyright reasons this diagram has replaced the diagram in the CCEA past paper

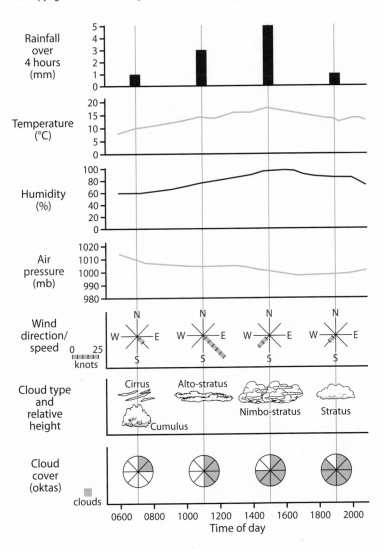

2. Using information from Resource B, explain the general weather conditions associated with this anticyclone. [6]

Question from CCEA AS1 Physical Geography paper, June 2015, © CCEA 2016

Resource B
Source: Data from the Met Office, Crown copyright 2016
Contains public sector information licensed under the Open Government Licence v1.0
For copyright reasons this diagram has replaced the diagram in the CCEA past paper

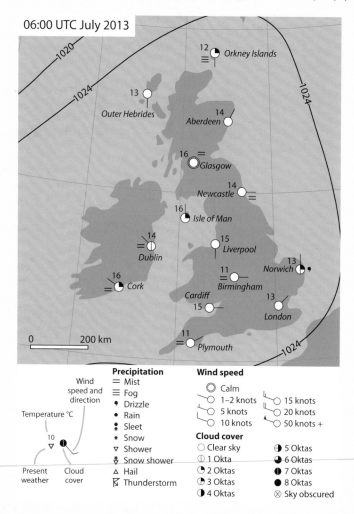

References

Geofile articles:
'A comparison of air masses affecting the UK', *Geofile* 730, series 33, 2014–2015
'Depressions: two storms compared', *Geofile* 703, series 32, 2013–2014
'Anticyclones', *Geofile* 552, series 26, 2007–2008
'Cold winters in the UK: a case study of 2012/13', *Geofile* 706, series 32, 2013–2014

Geo Factsheet:
'Depressions', *Geo Factsheet* 146, Curriculum Press
'Anticyclones – a potential hazard?', *Geo Factsheet* 106, Curriculum Press

Archive Met office educational material can be found at:
http://www.metlink.org/met-office-education-resources/

Other useful sites are:
www.sat.dundee.ac.uk
www.bbc.co.uk/weather

Unit AS 2:
Human Geography

1A Population data

Case studies

Contrasting national case studies

MEDC: UK

LEDC: Nigeria

Students should be able to:

(i) distinguish between:
- national census taking
- vital registration

(ii) demonstrate knowledge and understanding of the contrasts between MEDCs and LEDCs in relation to:
- the reliability of data
- how the data is collected
- the use made of the data

Population Geography studies the spatial variations in the growth, composition and distribution of the world's population. In particular, Population Geographers are concerned with the dynamics of population change both in terms of growth over time and space. In order for Geographers to describe and explain these variations there is a need for accurate and reliable data. Two main sources available to Geographers are:

- the national census
- vital registration

The national census

The word 'census' comes from a Latin word meaning 'tax' but it has come to refer to a count of all of the population and those social and economic characteristics that can easily be counted. The information collected provides a unique snapshot of the social, economic and demographic conditions of all of the people at a specific time. Censuses have been taken for many centuries, as governments have sought to collect information for many reasons. The ancient Egyptians collected information on their population numbers so that they could plan armies of people to build the Pyramids. The Romans carried out regular censuses throughout their empire in order to plan the regional distribution of power. In the *Bible*, it was one of these censuses ordered by Caesar Augustus that caused Mary and Joseph to travel to Bethlehem before the birth of Jesus. These early censuses were held irregularly, did not cover all of the population and were therefore inaccurate. Censuses take place at regular intervals in all MEDCs (usually every 10 years). In the UK a census has occurred every 10 years since 1801 with the exception of 1941. The last census in the UK was held on 27 March 2011.

Vital registration

Vital registration is the official recording of all births (including stillbirths), adoptions, marriage and civil partnerships, and deaths as they occur. Most countries have adopted this practice.

Vital registration serves two purposes:

1. In contrast to the census, which records information at a given point in time, vital registration records the events as they happen. Information on birth rates, death rates and causes of death can be obtained from these records. In this way, vital registration provides a continuous record of population change.

2. Vital registration is legally binding and is used to provide legal documentation for all individuals throughout their life. The data collected is held permanently and can be used to guarantee an individual's legal rights, such as nationality.

Data collection in MEDCs and LEDCs

The collection of accurate population data across an entire nation is fundamental to all censuses. This information provides crucial information to governments, enabling them to plan services for the future and to make informed decisions on the allocation of funding.

MEDCS

In MEDCs, accurate census-taking has largely been achieved. There are many reasons for this success, including:

1. There is a long history of census taking in MEDCs and the entire process is managed effectively. In the UK that task is the responsibility of The Office for National Statistics (ONS). This organisation is not affiliated to any political party so impartiality is assured.

2. A high level of literacy and helplines for those with literacy issues or visual impairment ensure that everyone can complete the census forms.

3. Completion of a census is now legally binding within the European Union and The United Nations strongly encourages all countries to complete a census.

4. The rationale behind the census is carefully explained through the media, with emphasis on the necessity of this information for future planning and the overall well being of the population as a whole.

5. Confidentiality of the information collected is assured. All of the information is processed in secure conditions and confidentiality is guaranteed by legal protection.

6. Every effort is made to ensure full participation in the census and there is a substantial fine imposed on anyone refusing to complete the census form.

7. The distribution of the census is carefully planned and every household in the country will receive its census package well in advance of the completion date. There are no major issues with distribution even to the most remote parts of MEDCs.

8. The information collected is processed and made accessible to all. Governments use this information when planning services such as schools, hospitals or care for the elderly. The UK government provides funding for the devolved parliaments in Northern Ireland, Scotland and Wales. The amount provided is calculated using information from the census.

9. Census questions and format are reviewed regularly to ensure that relevant information is collected. Some countries including the Netherlands, Austria, Switzerland, Israel and Singapore have partly or fully replaced the census with continuous analysis of information held by government agencies and sample

surveys. Currently, many MEDCs offer online census forms (see UK case study, pages 138–141).

10. Vital registration is legally binding in all MEDCs.

LEDCS

In LEDCs, census-taking has been much less successful in producing accurate and comprehensive information. A major issue is the infrequency of censuses, with many countries not having regular censuses. Often the data is unreliable or inaccurate and on occasions a census has been deemed unreliable. At other times a census has been cancelled. See Figure 1.

Figure 1: Census dates for LEDCs

Source: Data from UN Population and Vital Statistics Report (2007) and UN Statistics Division 2010 World Population and Housing Census Programme: Census dates for all countries. Updated from UN Statistics Division 2016 World Population and Housing Census Programme: Census dates for all countries (http://unstats.un.org).

Country	Year of previous census	Year of most recent census
Burundi	1990	2008
Uzbekistan	1989	2010
Somalia	1987	Scheduled for 2014. Cancelled.[1]
Dem Rep of Congo	1984	Scheduled 2010. Cancelled.[2]
Eritrea	1984	Scheduled for 2009. Cancelled.[1]
Myanmar	1983	2014
Togo	1981	2010
Afghanistan	1979	Scheduled 2008. Cancelled.[1]
Angola	1970	2014
Lebanon	1932	None taken
Liberia	1984	2008
Western Sahara	1970	None taken
Djibouti	1960	2009

[1] Cancelled for security reasons. [2] Cancelled for political reasons.

There are many reasons to account for these problems:

1. In some LEDCs their first experience of a census occurred during colonialism, when censuses were carried out by the colonial rulers for the purpose of tax collection, conscription to the army or simply to gain information about the number of their subjects. As such, the census came to be seen by the indigenous population as a symbol of foreign power and domination. The local population sometimes reacted by providing inaccurate information.

2. In some areas political power, parliamentary seats or budget allowances are determined by information gathered in a census. Groups within the population may attempt to gain political advantage in these areas by providing inaccurate information. During the 1941 census in pre-partition India there was a tendency for the two main ethnic groups – Hindus and Muslims – to exaggerate their

numbers because it was commonly believed that the country would be partitioned on ethnic lines. More recently, in the state of Kashmir, a boycott by some ethnic groups threatened to obstruct the operation of the 2001 census.

3. Gross per capita income is often used to determine the amount of international aid allocated to LEDCs and therefore a large population could be seen as an advantage. Several incidents of over-estimations of total population size have been reported across Africa. Gabon refused to accept the census results for its total population, which the government considered to be too small, even though it was ratified by the United Nations.

4. In many ethnically diverse countries there is often a delicate balance between the numbers in each group and the allocation of power, and this has led to manipulation of the census data in favour of the ruling group. In Lebanon no census has been taken since 1932 for fear of data-tampering and the resultant unrest that might follow between Muslim and Christian groups.

5. Census-taking is expensive – the 2011 census in the UK cost approximately £480 million. Central government considered such expenditure worthwhile, in order to gain accurate and comprehensive information. Few LEDCs can afford such large sums of money.

6. In some LEDCs literacy problems mean that many censuses are collected by door-to-door interviews. This can result in some people being omitted or inaccurately recorded. In certain cultures women may not be interviewed by men, who collect the census information, resulting in further misrepresentation.

7. Within some LEDCs the nature of the data collected is hotly debated. On some occasions the questions on ethnicity, religion or race have led to violence (see Nigeria case study, pages 142–144).

8. In many LEDCs transport and communications are poor, resulting in some people being omitted from the census altogether. In some LEDCs there are still nomadic people who are missed in the census.

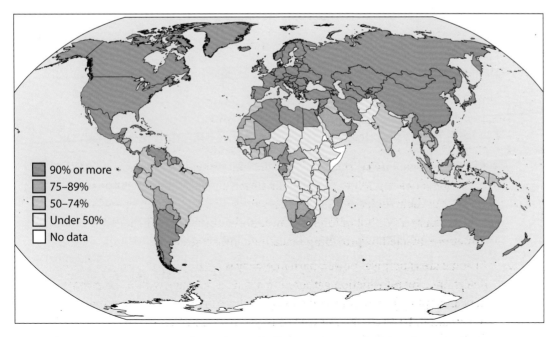

Figure 2: Coverage of birth registration

Data from United Nations Statistics Division, updated December 2014

*Figure 3:
Coverage
of death
registration*

Data from United
Nations Statistics
Division, updated
December 2014

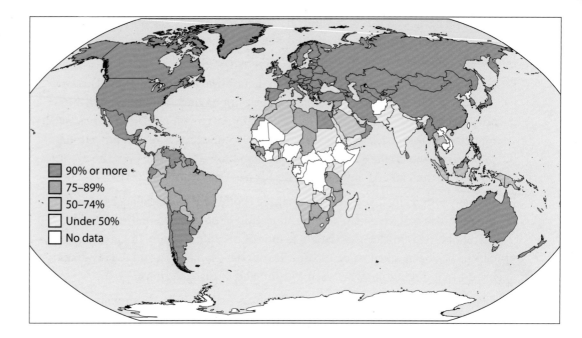

9.   In many LEDCs wars and civil unrest are common occurrences and this can result in a census being postponed, cancelled or incomplete (see Figure 1).

10. In some LEDCs census data is used to allocate funding or political power. This data is often controlled by the governing party, allowing it to be deliberately manipulated in the party's favour (see Nigeria case study, page 143).

11. In some LEDCs inadequate training of the census enumerators (collectors), lack of resources and administrative issues have led to incomplete census returns.

12. Since the 1950s the UN has supported most LEDCs in establishing vital registration practices. As Figures 2 and 3 show, most countries record births and deaths but African countries have the poorest record. According to the United Nations' data, many countries in Sub-Saharan Africa have under 50% of births registered (see Figure 2). A combination of literacy problems, poverty and poor infrastructure account for this.

Exercise

Describe the patterns of birth and death registration shown in Figures 2 and 3.

CASE STUDY: National population data collection in a MEDC, UK

The *Domesday Book*, produced during the reign of William the Conqueror, in 1086, was the first real attempt at conducting a national count of population, land and property in England. The main reason for collecting this data was to inform the new king of the extent of his conquests in England. There were other piecemeal attempts at population surveys from then on, but the idea of a national census to count all of the people did not come about until the end of the eighteenth century. This was a time of considerable social change in Britain – agriculture improved, industry developed, towns increased in size due to rural-urban migration and new advances in medicine resulted in increased knowledge about sanitation and nutrition. There was some debate about the impact of these changes on population numbers but most agreed that there was a significant increase in population size.

It was against this background of change and uncertainty about the future that Thomas Malthus published his theory on 'The Principle of Population' where he stated that population growth, if left unchecked, would soon exceed the available food supply and resources. In this situation, Britain would face famine, disease and war. It became increasingly clear that a national census was required to provide accurate information on population size. Initially there were objections, as some feared that a census could reveal too much information to the enemies of Britain, but parliament passed the Census Act in 1800 and the first census was carried out in Britain in 1801 and in Ireland in 1821. This first census was conducted by door-to-door collectors and it was not until 1841 that an official registrar nominated a specific date when each head of household would complete a form. This remained the format of the UK census until 2011.

Since its introduction in 1801, a census has taken place in the UK every ten years, with the exception of 1941 during the Second World War. The information collected provides a unique snapshot of the social, economic and demographic conditions of all of the people at a specific time. This information enables central and local government, and health and education authorities to target their resources more effectively and to plan health, education, housing and transport services for years to come.

The last census taken in the UK was in 2011 and already plans are well underway for a revised census format for 2021. In England and Wales the census is carried out and processed by the Office for National Statistics (ONS). In Northern Ireland, the Northern Ireland Statistics and Research Agency (NISRA) performs a similar function. For a census to be worthwhile it has to be reliable and fully inclusive. The UK, like most MEDCs, has succeeded in gathering accurate and comprehensive information.

Detailed planning is a key factor in the success of the UK census. Plans for the 2011 census included a pilot test of the data collection process in 2007 and a pilot test of the census form in 2009. The 2007 test introduced a number of additional questions to those used in the 2001 census. For the first time there were questions on:

- National identity – to allow respondents to record their English, Welsh, Scottish, Northern Irish, Irish or other identity.
- Income – to collect level and sources of income.
- Language – to collect information on proficiency in English, Welsh and other languages. In Wales, people were asked about the frequency of their use of the Welsh language.
- Second address – to identify the number of people with a regular second address and the purpose and frequency of its use.
- Month and year of entry into the UK – to collect extra information about international migration.
- Illness and disability – to collect information on the nature of illness and disability.
- Marital or civil partnership status – to include civil partnership equivalent for each marital status.

Following the test, a census rehearsal was carried out in 2009 and the revised census was submitted to Parliament and became legally binding later that year. An

Details of the 1801 census

Taken on 10th March 1801.

Details recorded for each parish, township, or place were:

- Number of inhabited houses, occupied by how many families
- Number of uninhabited houses
- How many persons, how many male, how many female
- How many persons are chiefly employed in agriculture; how many in trade, manufactures, or handicraft; and how many in neither
- How many baptisms and burials in the years 1700 to 1800, distinguishing males from females
- How many marriages in each year from 1754 to the end of 1800

Details of individuals and their names were not recorded in the official Census returns.

Source: 'The 1801 Census', http://www.1911census.org.uk/1801.htm

extensive media campaign was launched to explain the importance of the census and the penalties for non-completion.

The 2011 census was the first census that could be completed online. Households wishing to avail of this option were given a unique code which enabled their form to be accessed online. Approximately 15% of responses across the UK were online.

The overall return rate to the census was 95%. The data collected from the 2011 census was published from 2013 onwards and will be used to advise governments on service provision for the next ten years. The overall cost was calculated at £480 million.

Soon after the 2011 census, the UK Statistics Authority, the government body that has overall responsibility for the census, undertook a major review of the current methods of data collection. They set up the **Beyond 2011 Programme** to examine two areas of concern regarding the census process:

1. Although, the UK census has an excellent return rate, 95% in 2011, it is not universal.
2. In a rapidly changing world with large numbers of migrants, some critics suggest that a ten yearly cycle of data collection is inefficient. Government Agencies such as the NHS, the tax office and social services have vast amounts of information about the population which could be accessed and processed. It was suggested that the possibility of accessing and coordinating this information might make the need for a census redundant.

The Beyond 2011 Programme reported their findings in 2014. They concluded that the census should continue but the form should be largely completed online. For those who have no internet access, postal census forms will be available. It is also proposed that the census data will be supplemented through access to the existing information held by Government Agencies.

Vital registration

Vital registration has been compulsory in England and Wales since 1837 and in Northern Ireland since 1864. Births, still births, deaths, age and cause of death, marriage and civil partnerships are some of the data collected. All births including still births must be registered within 42 days of birth. Deaths must be registered before burial takes place. The legal documentation takes the form of birth certificates, death certificates and marriage certificates. In the UK an individual must produce a birth certificate to register with a doctor, to obtain a passport, to apply for a visa to live in a foreign country or to get married.

The management of the information collected is the responsibility of the General Records Office (GRO) in England and Wales. There are equivalent offices in Northern Ireland and Scotland. In this way, vital registration provides a continuous record of population change. In non-census years the total population can be estimated using the information held by these offices. For example, the total population for Northern Ireland in 2012 (one year after the census) was the total population recorded in the census, plus the new births, minus the deaths, plus or minus the net migration in that year (see Exam Question 4, page 144). The Office for National Statistics (ONS) frequently publish data from vital registration (see Figures 4 and 5 below).

Figure 4: Number of Live Births and Total Fertility Rate (TFR), 1943–2013 England and Wales

Source: 'Births in England and Wales: 2013', release date 16 July 2014, Office for National Statistics licensed under the Open Government Licence v.3.0.

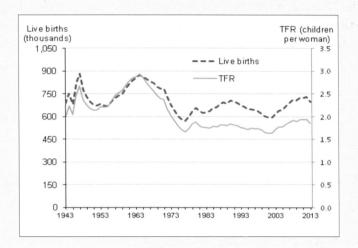

Figure 5: Infographic on birth statistics 2013

Source: 'Births in England and Wales: 2013', release date 16 July 2014, Office for National Statistics licensed under the Open Government Licence v.3.0.

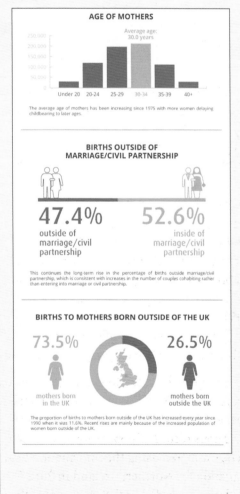

CASE STUDY: National population data collection in a LEDC, Nigeria

Nigeria, formerly part of the British Empire, became an independent country in 1961. The country has over 250 ethnic and linguistic groups largely grouped into a Muslim north and a Christian south. Nigeria is potentially Africa's richest country and since 2013 has been part of a group of newly emerging market economies (MINT, see pages 252–253). However, decades of mismanagement and political unrest have hindered its progress. It is also Africa's most populous country but the total population is unknown as there has never been a reliable census. Distrust between the two main regions has undermined every census attempt and this lack of reliable information is one reason for Nigeria's failure to utilise its resources adequately.

The first real attempt at a national census in Nigeria was in 1952–1953, when the country was still under British rule. There have been infrequent censuses since then with the next census postponed until 2017 due to lack of funds and logistical problems according to the Chairman of the National Population Commission (NPC, the organisation responsible for population data in Nigeria). As Figure 6 shows, none of these censuses has produced reliable data.

Figure 6: Nigerian Census Records 1952–2006

Data from 'The recurring controversy over national census – Nigerian Tribune', *The Citizen*, http://thecitizenng.com

Year	Population size (millions)	Official verdict on the reliability of the census
1952–53	31.6	Incomplete survey*
1962	45.26	Contested and a re-count ordered
1963 (re-count)	55.6	Unreliable
1973	83.0/79.8	Contested; revised; unreliable
1991	88.9	Unreliable
2006	140	Unreliable

*Parts of eastern Nigeria not surveyed due to political unrest

As is the case with many LEDCs, Nigeria faces various obstacles to gathering accurate and reliable data. The first census in Nigeria was conducted under British rule and many Nigerians were suspicious that it was related to taxation. This suspicion is thought to underlie the alleged undercount at this time. Political unrest has frequently occurred during census taking in Nigeria and such unrest resulted in large parts of eastern Nigeria being omitted from the 1952–1953 census.

Parts of the country have difficult terrain and poor infrastructure and, as the Nigerian census is conducted using door-to-door interviews, some communities are omitted. There is often inadequate preparation for the census count. During the 2006 census people were ordered to stay at home on the day appointed for the count in their area. However, there were insufficient forms at the start of the census, which resulted in some areas missing out. Furthermore, as many as 45,000 nomadic people were thought to be have been left out because they had not been included on the official census map. Thousands of enumerators abandoned their work, complaining of insufficient funds to carry out the census.

In northern Nigeria census enumerators are only allowed to interview the head of household. This means that the accuracy of responses cannot be verified. In a country with over 250 linguistic groups there is inadequate help given to overcome language barriers.

One of the greatest stumbling blocks to obtaining reliable census data in Nigeria relates to the government practice of allocating funds and political power to population size. Therefore, individual states have a major financial interest in their total population numbers. It is claimed that the 1952–1953 census manipulated the figures in favour of the northern states in order to allocate them more seats in parliament and to diminish the influence of the southern states. The first post-colonial census in 1963, with a population total of 55.6 million, would have needed an average annual population increase of over 5% and in some southern states an annual growth rate of 13%. These figures were so unrealistic that the census was deemed a failure. The census of 1973 was similarly lacking in credibility and was also deemed a failure.

Closely linked to these problems of data collection is the nature of the data collected. The early censuses did not collect information on the religious or ethnic identity of the population. In an attempt to improve accuracy of population figures it was proposed that ethnic and religious identity questions would be included in the 2006 census. Fearful that the inclusion of these questions might diminish their political advantage, the Northern states threatened to boycott the census if such information was requested, while the southern states threatened a boycott if the questions on ethnic and religious identity were removed. In the end, the questions were removed but the census was seen as unreliable.

A census is planned for 2017 and there are many attempts to make this one reliable. There has been a media campaign to engage the people in the importance of gaining accurate information. Literacy levels are improving and there is greater access to media. The need to use the data for the overall benefit of the country rather than political favouritism has attracted considerable media coverage. Nigeria has seen an increase in its standard of living since 2006. There have been improvements in transport and infrastructure. The NPC has campaigned for adequate funding for proper training for the enumerators and for computing equipment to analyse the data. They also want to extend the use of biometric data. This would help overcome fraudulent counting. However, the cost, estimated to be around £640 million, is considerable even for a potentially rich LEDC. Furthermore, the NPC has limited influence over census regulation in Nigeria and a former chairman was forced to resign his post following his criticism of government intervention.

What is the population of Nigeria?

To date the most accurate information on Nigeria's population comes not from the census but from e-Geopolis, an organisation that provides population estimates on settlements over 10,000 people using a combination of satellite imagery and GIS. The Organisation for Economic Development (OECD) has used the information supplied by e-Geoplis and estimated the total population in Nigeria to be 134 million in 2006, rising to 162 million in 2016 if rates of population change remain constant.

Vital registration

Vital registration has been made compulsory in Nigeria since 1970 but it was not until 1990 that a national policy was developed under the control of the NPC. The NPC has produced data on live births, still births and age and cause of death. However, data collected by UNICEF indicates that there are still considerable variations between rural and urban areas in terms of the completeness of vital registration of births.

Exam Questions

1. Distinguish between vital registration and national census taking and discuss why population data collection is more effective in some countries than others. [6]
   Question from CCEA AS2 Human Geography paper, January 2012, © CCEA 2016

2. Describe two problems associated with population data collection in LEDCs. [4]
   Question from CCEA AS2 Human Geography paper, June 2012, © CCEA 2016

3. "Population data collection in MEDCs is always reliable and accurate. This is not the case in LEDCs." With reference to one case study of population data collection from a MEDC and one from a LEDC, discuss the extent to which you agree with this statement. [12]

4. Study Resource A below which shows population change in the UK between 2011 and 2012. Describe the relative contribution of natural population change and migration to population change in the UK as illustrated by these figures. [4]

Resource A

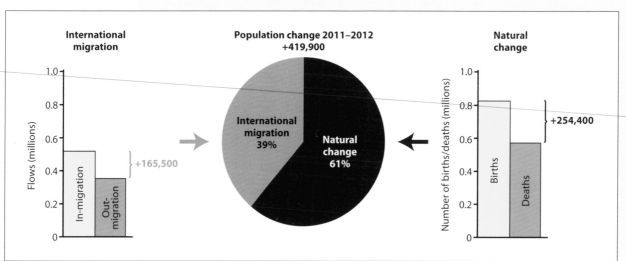

Source: Adapted from the 'Annual Mid-year Population Estimates, 2011 and 2012', 8 August 2013. Adapted from data from the Office for National Statistics licensed under the Open Government Licence v.3.0.

References

Plans for the 2021 census in Northern Ireland:
www.nisra.gov.uk/archive/census/2021/planning/the-future-provision-of-census-of-population-information-for-northern-ireland.pdf

Information about the 2011 census in Northern Ireland:
http://www.nisra.gov.uk/census/2011Census.html

The 2011 census form for Northern Ireland is available from the following website link:
http://census.ukdataservice.ac.uk/media/50969/2011_northernireland_household.pdf

Information about vital registration in Northern Ireland:
http://www.nisra.gov.uk/demography/default.asp2.htm

The Office for National Statistics (ONS) website includes 2011 census and vital registration data:
www.ons.gov.uk/

Information on future data collection methods in the UK:
http://www.ons.gov.uk/ons/about-ons/who-ons-are/programmes-and-projects/beyond-2011/index.html

Information on the Nigerian census is available from the following website links:
http://www.prb.org/Publications/Articles/2007/ResultsFromNigerianCensus.aspx
http://www.prb.org/Publications/Articles/2007/ObjectionsOverNigerianCensus.aspx
http://www.prb.org/Publications/Articles/2006/IntheNewsTheNigerianCensus.aspx
https://africacheck.org/factsheets/factsheet-nigerias-population-figures/
http://www.informationng.com/tag/national-population-commission

A copy of the Nigerian 2006 Population and Housing Census can be found on the following website link:
http://catalog.ihsn.org/index.php/catalog/3340

Case studies

General reference
to places for
illustration
purposes only

Students should be able to:

(i)   describe the main fertility and mortality measures – crude birth rate, crude death rate, total fertility rate and infant mortality rate

(ii) demonstrate knowledge and understanding of the demographic transition model and the epidemiological transition

As the previous section highlighted, accurate population numbers are extremely important and are a key reason for carrying out a census. According to a United Nations Report in 2015, the world population increases by approximately 83 million annually. Currently more than half of this occurs in Africa and Asia.

Changes in population numbers are largely due to variations in the numbers of births and deaths. The balance between the number of births and the number of deaths is known as natural population change. Natural population change is calculated through a number of fertility (birth) measures and mortality (death) measures.

*Figure 7: Most
populous countries
2015 and 2050
(projected)*

Source: '2015 World
Population Data Sheet',
Population Reference
Bureau, www.prb.org

2015		2050	
Country	**Population (millions)**	**Country**	**Population (millions)**
China	1,375	India	1,660
India	1,314	China	1,366
United States	321	United States	398
Indonesia	256	Nigeria	397
Brazil	205	Indonesia	366
Pakistan	199	Pakistan	344
Nigeria	182	Brazil	226
Bangladesh	160	Bangladesh	202
Russia	144	Congo, Dem Rep	194
Mexico	127	Ethiopia	165

Fertility measures

Crude Birth Rate (CBR)

CBR is the total number of live births per thousand of the population per year. This measure has several weaknesses, largely because it does not take account of the age or gender composition of the population. However, it is simple and widely used.

There is a clearly defined global pattern of CBR. In MEDCs, CBR is typically between 9–12 per thousand. MEDCs are rich, industrialised countries with universal access to education and health care. The material benefits of small families are well understood and contraception is widely available. Gender equality enables women to pursue careers and the age of marriage has increased. All of these factors underlie a low birth rate. The knock-on effect of low birth rates results in ageing of the population and fewer potential parents, thereby causing CBR to decrease further.

Overpopulation

At one end of the spectrum there are too many people for the existing resources to be used in a sustainable way. Resources are overused or depleted. Standard of living falls, eventually leading to extreme poverty.

Underpopulation

At the other end of the spectrum, there are insufficient numbers to utilise the resources efficiently. In this situation an area is not developing its economic potential fully.

It is important to remember that the balance between population and resources is not static but changes over time. A finite resource in area may be depleted and fewer people can be sustained. If new resources are not available the area may move from a situation of optimum population to one of overpopulation.

Figure 18: Optimum population graph

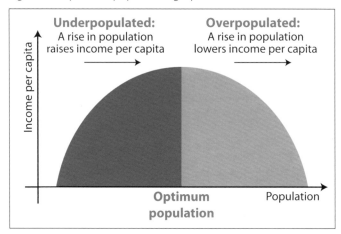

Theories and issues of population sustainability

Issues concerning population and sustainability are frequently reported in the media and world leaders meet to discuss ways of sharing resources more equitably between MEDCs and LEDCs. However, these concerns over resources and sustainability are not recent. Just over 200 years ago (1798) **Thomas Malthus** produced his work on 'The Principle of Population Growth'. In this, he stated his concerns that population growth, if left unchecked, would outstrip the available resources. In this context, a resource referred to food supply. The basis of his theory rested on the differential rates of growth for population and resources. He claimed that human populations increased at a much faster geometric rate (1, 2, 4, 8, 16) and had the potential to double every 25 years. Food supplies on the other hand only increased at an arithmetic rate (1, 2, 3, 4). Based on this belief, Malthus claimed that there were a finite number of people that a country could sustain and if population growth was not reduced then certain environmental checks would come into play. In this way, population numbers would be reduced and a balance between population numbers and available resources would be re-established. Malthus states that the balance between population numbers and food resources can be regulated in two main ways:

1. Preventative (negative) checks

These involved delayed age of marriage and abstinence from sex within marriage. This would lower the birth rate. It should be noted that Malthus was writing before contraception was available.

2. Positive checks

These referred to changes in the death rate. Malthus believed that when there was insufficient food available, disease, famine and war were likely outcomes, resulting in an increase in the death rate.

Two concepts are fundamental to Malthus' views. These are:

- carrying capacity – which is the maximum capability of a region to support people with food.
- population ceiling – which is the numerical limit of people who can be supported in any given region.

Figure 19: Malthus' scenarios of population and resource imbalance

In his writing, Malthus proposed three possible scenarios when population and resources were imbalanced (see Figure 19).

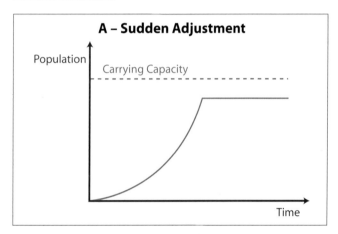

In diagram A – the population grows unchecked until the carrying capacity is reached. When the carrying capacity is reached, population growth stops and remains at the carrying capacity level. There is no evidence that such an example has occurred.

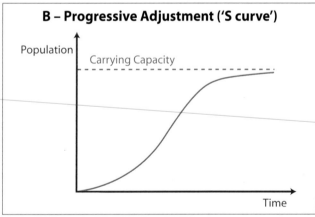

In diagram B – the population growth begins to slow down before the carrying capacity is reached and levels off close to the population ceiling. This is said to be typical of regions that are in control of their resources and appreciate the problems ahead, and so take decisive action beforehand. Population and resources are kept in balance by changes to the fertility levels – the so-called preventative checks.

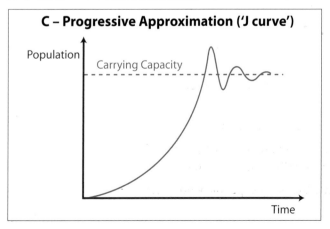

In diagram C – the population growth rises beyond the carrying capacity of the region and then numbers are reduced either through the positive checks of famine, war and drought or through the negative checks of birth control and delayed marriage. This process continues for a time before population levels off close to the carrying capacity.

Malthus' theories have been widely debated and as yet no agreement has been reached on the viability of his views in our world today. In 1965, nearly 200 years after Malthus, **Esther Boserup** proposed a very different view concerning the balance between population and resources. Boserup claimed that population growth was a necessary prerequisite for technological advances and innovation, and that society was held back by slow population growth. She believed that populations would continue to grow until they came close to the carrying capacity and at that point human inventiveness would find a way to avert the crisis and produce more food. This is an optimistic viewpoint but there is evidence to support it. Boserup claimed that population was increasing in Britain during the eighteenth century. At this time agricultural reform resulted in an increase in food production. Boserup argued that the increase in population was necessary for this reform to take place. From there, Britain's food production increased and the surplus was sufficient to feed the population working and living in the towns and cities that developed throughout the nineteenth century.

There are arguments for and against both of these theories.

Support for Malthus/against Boserup

- There are many examples of famine which have been partly caused by overpopulation, such as the potato famine in Ireland in mid-nineteenth century.
- Desertification in the Sahel region in Africa is partly due to overgrazing on semi-arid land. The Ethiopian famine in the 1980s was partly due to desertification.
- In the Middle East, conflict between Israel, and Syria and Jordan has been exacerbated by rival claims for scarce water resources from the river Jordan.
- Boserup believes population will always know how to get more food from the land but this is too simplistic.

Against Malthus/support for Boserup

- There have been dramatic improvements in agriculture. By using modern methods, such as chemical fertiliser, high yielding crops and genetically modified crops, we have increased food output.
- Countries can import food so there is no longer a need to be self-sufficient in food.
- More land has been brought into production and GM crops can be grown on land previously considered unsuitable.
- The Green Revolution in the 1950s produced new high-yielding cereal crops that have had considerable success in many Asian countries.
- Many Asian countries are able to grow two crops a year, thus doubling the output from their land.
- Increases in the distance that fishing boats can travel has led to more species being available.
- The use of irrigation means semi-arid lands can be made productive.
- Food mountains and food waste are used as evidence that there is adequate food available, and that scarcity of food is a problem of distribution not supply.
- In countries experiencing famine it is usually only the poorest people who suffer from lack of food.

- The problems of desertification are thought to be due, in part, to inappropriate farming techniques.
- Wars and corrupt governments often cause food shortages, either through destruction of crops or in some cases food is withheld from rival groups. During the long civil war in Ethiopia, the government was frequently accused of withholding food supplies from the rebel territories in Eritrea.

These are just some of the arguments for and against Malthus and Boserup. More recently a number of debates have taken place concerning this issue. The attention now seems to have moved away from the Domesday scenario of lack of food, although there are still areas where food supply is very inadequate, and focuses on the wider environmental implications of overconsumption of resources. Modern day protagonists of Malthusian views include Paul Erlich and the Club of Rome. They emphasise the environmental impacts of excessive population growth, including global warming and ozone damage. The views of Ester Boserup were championed by Julian Simon, who believed population growth was essential to increased economic prosperity.

Exam Questions

1. How do the theories of Thomas Malthus and Ester Boserup differ regarding the concept of population sustainability? [6]

Question from CCEA A21 Human Interactions and Global Issues paper, January 2011, © CCEA 2016

Exercise

Using information from Resources A and B, **and** from any two of the websites listed in the 'References' (page 165), discuss the extent to which you agree with the statement:

"There is no such thing as overpopulation. Malthus' ideas are outdated."

Resource A: Hydroponics

In nature, soil is essential for plants. It supports the plants, and delivers nutrients and moisture to them. However, nutrients and water can always be delivered without soil. Hydroponics, from the Greek 'hydro' (water) and 'ponos' (labour), is the term used for growing plants in mineral solutions, rather than in soil. It is used for growing vegetables, fruit, forest seedlings, etc, often in greenhouses.

It offers many advantages over soil-based cultivation:

- Water consumption is reduced, as the water stays in the system and can be reused. Hydroponics is said to consume only 5% of the water consumed by traditional farming.
- It can lower costs, as water consumption is reduced and nutrients can be controlled very carefully.
- Chemical run-off into groundwater and nutrient pollution are both reduced because the nutrients are more controlled and mineral solutions are contained within a closed system.

economic competitiveness. Many road projects have been undertaken to deal with this increase. A major Ring Road skirts the outer edge of Belfast with links to the M1 and M2. Bypasses have been built around several towns, including Holywood, Comber and Hillsborough. Other roads have been widened to add additional lanes, such as the A2 from Belfast to Carrickfergus. Often these developments have taken over farmland. Occasionally where the planned road developments would encroach on land of scenic or recreational value there has been considerable public protest. A notable example of successful opposition occurred over plans to build a road through part of Belvoir Park Forest in south Belfast, cancelled in 1995. More recently planners have become aware of the pollution caused by increased traffic and attempts have been made to make other forms of transport more attractive through the use of bus lanes and bicycle lanes.

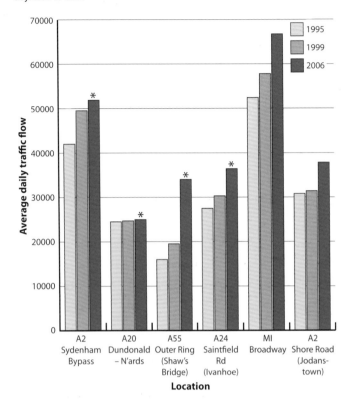

Figure 28: Traffic growth in the Belfast Metropolitan Area 1995–2006
Source: 1995 and 1999 - Annual Traffic Census Report 1999, 2006 - DRD
*adjusted to 2006

Another major issue is the building of retail centres in the rural-urban fringe. Greenfield developments such as Sprucefield Centre (close to Lisburn) and Abbeycentre (in Newtownabbey, close to Belfast) have attracted criticism from local residents because of the increased traffic, and from local town centre retailers for the loss of income. Once established these centres also attract additional retailers. A plan by the John Lewis organisation to build a store at the Sprucefield Centre was successfully opposed by local residents in 2013.

Urban areas generate a large amount of waste which has to be collected and managed by local councils. Some of this waste goes to landfill sites and the location of such sites is nearly always controversial. This is especially the case when the chosen site is in the rural-urban fringe. A major landfill site is located on Belfast's rural-urban fringe close to Lisburn. This site has attracted opposition from local residents, who claim it poses a health hazard and has a negative impact on house prices.

(2) Suburbanisation

Urban areas increase in size through urban sprawl or **suburbanisation**, a process that refers to the decentralisation of people, services and industry to the edge of the existing urban area. Cities grow outward from the centre in a series of stages and with each outward advance some rural environments are transformed into urban areas. In the case of Belfast, the first phase of suburbanisation occurred as early as the late 1920s and 1930s, when wealthy merchants and factory owners moved away from the densely populated and unhealthy inner parts of Belfast to the slightly higher and better drained lands at Stranmillis, Malone and Ormeau in the south, Stormont in the east and along the Antrim Road in the north. This marked the first separation between place of work and residence. Suburbanisation has always depended on the ability of suburban dwellers to access their workplace with relative ease. For Belfast's earliest suburban dwellers that meant living close to the new public transport routes of trams and railways. The outer limit of suburbanisation was determined by the extent of the public transport network. Over the next decades the process of suburbanisation continued to expand as public transport improved, but large-scale suburbanisation did not occur until private transport became generally available.

During the 1950s and 1960s two major developments led to rapid suburbanisation and to the first serious challenge for the surrounding rural environments:

(i)   At this stage, the condition of many of the working class housing areas had deteriorated to an unacceptable level and some people were re-housed in public sector housing at the edge of the city, in what were then greenfield sites. Examples of such public housing can be found in Castlereagh, Cregagh and Sydenham.

(ii) There was a general increase in the standard of living in the 1950s and people became more affluent, which enabled many to buy into the private housing market. Suburban housing developments continued to be built at the edge of the city.

The result of these developments was a sprawling city with more and more people leaving the inner areas in favour of the more attractive suburbs. In a reversal of previous trends, new industry followed the people to the purpose-built industrial estates, such as Castlereagh Industrial Estate surrounded by the public housing estates of Belvoir, Cregagh and Ballybeen. This in turn was followed by retail suburbanisation. The present retail site of Forestside in south Belfast was formerly occupied by Supermac, one of the earliest edge-of-city shopping complexes. It was built in 1964 on what was then a greenfield site.

Figure 29: Public sector housing in part of the Belvoir Estate, Belfast

Figure 30: Map of Belfast and suburbs

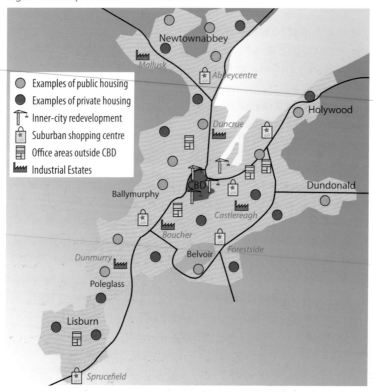

(a)

(b)

(3) Counterurbanisation

It appeared that the process of suburbanisation was now posing a significant threat to the surrounding countryside. Former villages such as Newtownbreda were absorbed by the growing suburbs. The Matthew Stopline (see 'Greenfield developments', page 176) was successful to an extent. Suburbanisation was curtailed for a while but after considerable pressure, permission was eventually granted to build public sector housing across the stopline in the west of the city at Poleglass. Once the stopline had been breached for public housing, the private sector mounted increased pressure for the release of land in the greenbelt for private housing. Eventually land was released to a number of private developers, notably at Cairnshill and Four Winds in the south east.

The Belfast Urban Area Plan 2001 took additional steps to prevent further suburbanisation, but if these containment policies prevented urban sprawl, another process was challenging the rural areas. This was counterurbanisation – a movement of urban workers to rural towns and villages within commuting distance of the city. Towns such as Comber, Carryduff, Saintfield and Hillsborough have all experienced considerable growth of private housing in greenfield sites in the last 25 years. It is not just the increase in population that challenges these rural towns but the character of the towns is often changed significantly. The new urban migrants often have little connection with their adopted living place. They work, shop and use the leisure facilities of the urban area. House prices rise as housing in towns that allow a rural living place close to an urban work place becomes highly desirable. This can be an advantage for the original inhabitants if they wish to sell property or land for development but a considerable disadvantage to the rural dweller who wants to buy into the housing market.

Despite such opposition to new developments in greenfield sites, the process of counterurbanisation and the challenges it poses for rural areas will continue and planners have to formulate policies to deal with the increased demands for housing in the rural-urban fringe. In 2006, Belfast Metropolitan Area Plan (BMAP) was published. This outlined future strategies and demands in the commuting hinterland of Belfast, including the rural-urban fringe. It was estimated that an additional 51,000 houses will be required in the Belfast Metropolitan Area (BMA), 9,000 of which will be built on greenfield sites. In 2015 this was revised to 66,500. A major issue for the planners is how they can reconcile this need against their stated aim to maintain the rural character of the settlements in the outer parts of the BMA through very strict planning controls on new buildings. Most of the 9,000 new homes will be in designated settlements including Carryduff, Ballyclare and Moira. These new housing developments on greenfield sites will be subject to strict planning policies.

Figure 31:
(a) Original housing (modernised) within Newtownbreda village and
(b) modern housing on the outskirts

The intention is to create balanced communities by building a mix of housing tenures including owner-occupied, public sector or social housing and specialised housing for the elderly and people with disabilities. In this way they hope to prevent the building of uniform housing developments that characterised the earlier commuter settlements.

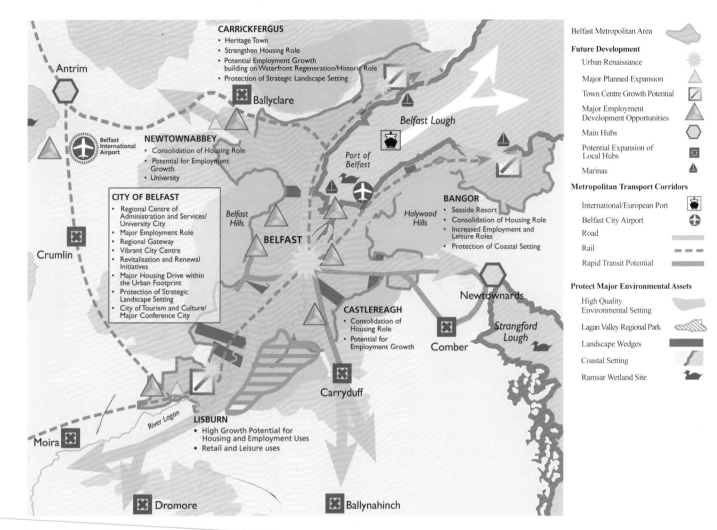

Figure 32: Planning Map for the Belfast Metropolitan Area

Source: 'Shaping Our Future: Regional Development Strategy for Northern Ireland 2025', Department for Regional Development, licensed under the Open Government Licence v3.0.

Exercise

1. Study the information in Resource A.

    (a) What were the planners hoping to achieve by these recommendations?

    (b) Suggest one group of people who might object to the creation of a Metropolitan Development Limit around existing settlements in greenfield sites.

2. Study Resource B which shows changes in the population of the council areas in the Belfast Metropolitan Area 1971–2011.

    (a) Describe the patterns of population change illustrated.

    (b) Identify and explain the processes that might have brought about these changes.

Resource A

In September 2014, the environment minister at Stormont accepted the recommendations of BMAP 2015. Although the plan has not yet become law some of its recommendations include:

- to designate a Metropolitan Development Limit around existing settlements in greenfield sites. Planning permission will not be granted for any building outside this limit.
- to direct new housing as far as possible to brownfield and sites within Belfast.
- to encourage new industry within Belfast eg in Titanic Quarter, Harbour Exchange.
- improving the attractiveness of Belfast City both as a place of work and residence.
- improving public transport, including the use of Park and Ride facilities in the suburbs.

Resource B: Population of council areas in the Belfast Metropolitan Area 1971–2011

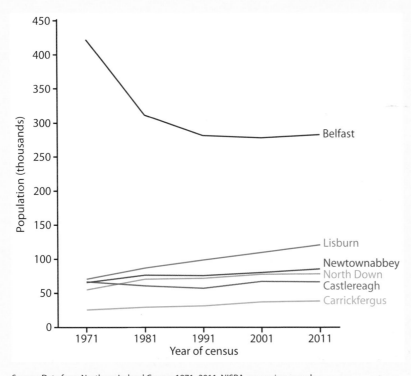

Source: Data from Northern Ireland Census 1971–2011, NISRA, www.nisra.gov.uk

Reference
Additional information on BMAP2015:
http://www.planningni.gov.uk/index/policy/dev_plans/devplans_az/bmap_2015.htm

Exam Questions

1. With reference to place for illustration, discuss three issues faced in the rural-urban fringe. [12]

Question from CCEA AS2 Human Geography paper, January 2012, © CCEA 2016

2. Study Resource I below which shows the percentage of the population living in metropolitan areas by state in the United States of America (USA) in 1910, 1950 and 2000 and answer the questions which follow.

(a) Describe the trend from 1910–2000 in the USA shown in Resource I. [2]

(b) Name the urban process shown in Resource I. [1]

(c) Describe one possible effect of this process on the rural-urban fringe of metropolitan areas. [3]

Question from CCEA AS2 Human Geography paper, June 2013, © CCEA 2016

Resource I

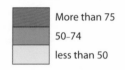

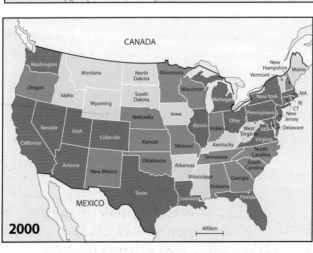

socio-economic reasons for this clustering, such as a feeling of safety, friendship and access to ethnic services and affordable housing, but it has also led to a 'ghetto' mentality, especially if the immigrants maintain their separateness. This was often emphasised by the out-migration of the white population from the streets close to an immigrant cluster.

Social and economic deprivation

It is easy to see that such areas today do not attract much investment and the jobs that are available are low paid. With little money and often poor educational attainment, many of the younger people feel alienated and see little opportunity of escaping this cycle of deprivation. Many of these areas are associated with anti-social behaviour such as drug-taking, crime and gang culture. Individuals become part of a self-perpetuating cycle of poverty (Figure 38). In such circumstances, the cycle of poverty is passed on from one generation to another. Children from disadvantaged homes are more likely to do poorly at school and have few qualifications, which reduces their opportunities for economic improvement. They are unable to secure a well-paid job and so they become victims of multiple deprivation. Several British cities, including London, Liverpool and Manchester, experienced riots in the early 1980s. Whilst the riots were for the most part clashes between black youths and the police, they highlighted the social and economic deprivation of these areas. Belfast had its own particular political problems, but much of the paramilitary activity emanated from the inner city areas.

Figure 38: The cycle of poverty

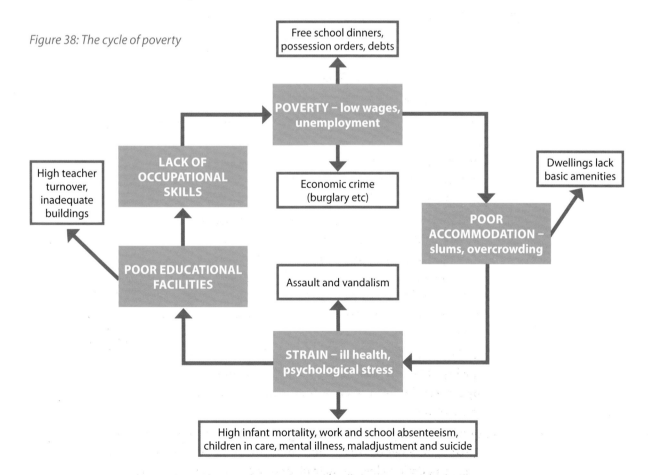

Alongside the problems of social and economic deprivation, inner cities were also affected by visual and land pollution. Often factory buildings were left derelict, adding further to the sense of decline that pervaded much of these areas. Some of the land was also polluted from earlier industrial usage and required expensive decontamination before it could be reused (for example, The Gasworks in Belfast). Such derelict sites (brownfield sites) had little attraction for modern industry. Increasingly, modern industry was attracted to edge of city locations (greenfield sites) in purpose-built industrial estates close to a labour supply and good communications. Figure 39 compares an inner ward of North Belfast with the rest of Belfast and Northern Ireland, using figures from the 2011 census.

Figure 39: Social and economic comparisons of an inner city ward (North Belfast) with Belfast and Northern Ireland, 2011 Census

Source: Data from Northern Ireland Census 2011, NISRA, www.nisra.gov.uk

Indicator	North Belfast	Belfast	Northern Ireland
Households rented (%)	69	45	30
Households with car ownership (%)	32	62	67
With no/low level qualifications* (%)	64	41	40
With university degree (%)	11	20	24
Unemployed (%)	9	6	5
Single parent families (%)	18	12	9
Births to un-married mothers (%)	75	58	43
Life expectancy (years)	73	79	80
Potential years of life lost (PYLL)** (2008–2012)	12	8.7	8.9

*low level qualification means 1–4 GCSE or equivalent at any grade.
**Years of life lost due to premature death compared to predicted life expectancy.

Exercise

1. Describe and suggest reasons for the patterns shown in Figure 39.

2. The use of Geographic Information Systems (GIS) allows geographers to investigate issues such as levels of deprivation within Belfast. The Northern Ireland Statistics Research Agency (NISRA) publishes the Northern Ireland Multiple Deprivation Measure (NIMDM). This measure is made up of 52 different indicators, grouped into seven major categories, which are used to describe levels of deprivation. The seven categories are:

   - income deprivation
   - employment deprivation
   - health and disability
   - education, skills and training
   - proximity to services
   - living environment
   - crime and disorder

Districts are ranked according to their combined score on these seven categories. Low scoring districts are more deprived than high scoring districts.

Use the following instructions to investigate levels of deprivation within Belfast:

   1. Open the following website link: http://www.ninis2.nisra.gov.uk/public/InteractiveMapTheme.aspx?themeNumber=137

All of the dockland areas have developed extensive modern restaurants, cafés and bars. The canals have been developed as amenity areas with various types of water-based activities, especially along Grand Canal Dock. New luxury apartments have been constructed in each development. A new road bridge (Samuel Beckett Bridge) and a footbridge link developments at the IFSC and North Lotts with the docklands south of the river.

Figure 47: Water based amenity at Grand Canal Dock

The banking crisis, which led to the economic downturn in 2008, had a severely negative impact on Ireland as a whole but the construction industry was worst affected. Many of the proposed developments in North Lotts and Grand Canal Dock were shelved as developers faced bankruptcy. In 2012, the government disbanded the DDDA and set up Strategic Development Zones (SDZ) to complete the regeneration of these dockland areas. Many new development programmes are currently underway.

This example clearly shows how the Dublin Docklands area has been gentrified. An area that was previously dominated by a working class population, living in low amenity, local authority housing with no attempt at landscaping, has been replaced by an affluent population, living in high quality and well designed accommodation, within walking distance from their place of work and recreational amenities.

Dublin Inner City Partnership (DICP), a body set up to study inner city poverty, compared the social and economic characteristics of this area between the 1991 census (before the re-urbanisation) and the 2006 census (10 years after the re-urbanisation). Some of their findings are shown in the table below. According to DCIP, these changes are typical of the gentrification process.

Percentage	1991	2006
Population change	−13 (1986–1991)	+80 (1991–2006)
Dependency ratio	43	16
No qualifications	74	20
Higher education/university degree	10	46
Male unemployment	63	14
Female unemployment	48	13
Local Authority housing	74	27
Private rental accommodation	8	52*
Owner occupied housing	16	25

Figure 48: Social and Economic Changes in North Dock C Ward (area of CHDDA) between 1991 and 2006 (before and after re-urbanisation)

Source: Data from Haase, Trutz, 'The Changing Face of Dublin's Inner City', a study commissioned by The Dublin Inner City Partnership.

* Many apartments in the IFSC district are for private rental

There is no doubt that the inner city wards of Dublin Docklands have been transformed in the last 25 years. It is now a fashionable, prosperous district with impressive architectural buildings, bridges, well designed open spaces and excellent communications. The economy of the region is now based on financial and digital services, which require a skilled workforce. Salaries are high and the employees have considerable disposable incomes. The leisure and recreational amenities in the area serve the needs of the new population.

Critics of the development in Dublin Docklands point to the gentrification that has occurred and the social segregation that pervades much of the area, even though the building of gated communities is no longer permitted. The affordable/social housing schemes provided few opportunities to those on low incomes and areas such as Sherriff Street still display evidence of economic and social deprivation. Approximately 120 people from Sherriff Street found temporary employment during the construction of the IFSC. The residents of these deprived areas do not have the skills required for the jobs currently available in the IFSC or Silicon Docks. Until this skills deficit is addressed, areas such as Sherriff Street will remain areas of high unemployment.

Exam Questions

1. With reference to your MEDC urban case study of an inner city:
   - explain why social and economic deprivation has occurred.
   - describe the attempts that have been made to improve this situation. [15]

2. (a) Explain the meaning of re-urbanisation and gentrification. [4]
   (b) With reference to your urban case study, explain why these processes can often occur in inner cities in MEDCs. [15]

References

Additional information on regeneration in Dublin:

Trutz Haase, 'Divided City, The Changing Face of Dublin's Inner City' – *http://trutzhaase.eu/wp/wp-content/uploads/R_2009_Divided-City.pdf*

Dublin Docklands – *www.dublindocklands.ie/*

Geofile articles:

'Case Study an Urban Centre Undergoing Redev'ment: West Bromwich', *Geofile* 698, series 32, 2013–2014

'Planning Issues in Today's MEDC Cities', *Geofile* 675, series 31, 2012–2013

Issues and challenges in LEDC cities

According to a United Nations report on Human Settlements 2016, 54% of the world's population now lives in urban areas. By 2050, the proportion living in urban areas is predicted to rise to 66%, approximately 6 billion people. Growth in urbanisation displays marked global variations. Although MEDCs are the most urbanised countries, the rate of urbanisation is greatest in LEDCs, especially those in Africa and Asia. Currently, Africa and Asia have 40% and 48% of their respective populations living in urban areas but they are predicted to reach 56% and 64% respectively by 2050. A UN report claims that globally there will be an additional 2.5 billion more urban dwellers between 2015–2050, 90% of whom will be in Africa and Asia. Rapid urbanisation has been one of the most pressing issues facing practically all LEDCs in the last 60 years and there is little sign that the problems are diminishing. The problems of rapid urbanisation are exacerbated by the growth of mega cities (cities with a population greater than 10 million). In 1950 only New York was in this category. In 2015 there were 28 in total, mostly in LEDCs.

Urbanisation in LEDCs

The current situation in LEDCs shows marked contrasts to the urbanisation phase that occurred in MEDCs in the nineteenth and early twentieth centuries. Then large numbers migrated to the towns and cities to take advantage of the new jobs in the factories and associated service industries. In other words, urbanisation was both a result and a cause of economic development. It took place over a period of about 150 years, during which time society experienced radical social and economic change. In LEDCs, urbanisation is a relatively recent occurrence. More importantly, the underlying causes are as much to do with rural poverty as urban prosperity. Medical advances have resulted in a falling death rate in most regions and, although fertility levels are also falling, the rate of decrease has been less than was experienced in MEDCs. In many cases, population in LEDCs continues to increase because of the youthful population structure, putting greater pressure on already scarce rural resources and limited employment opportunities. Faced with extreme poverty, many rural dwellers believe that moving to the cities is their only chance of securing a better future. In parts of Sub-Saharan Africa, war and natural disasters (such as droughts and famines) are the primary causes of this rural to urban migration.

High rates of natural increase contribute to the growth in the size of LEDC cities. A recent UN report states that natural increase accounts for 60% of urban growth in LEDCs (whereas in MEDCs rural to urban migration was the main cause). This means that these cities have a substantial youth dependency and significant potential for future growth.

Growth of informal settlements

In all LEDC cities, the provision of the basic essentials for life is a major challenge. Consequently, many of the poorest people do not have access to the most fundamental requirements for life, such as shelter and clean water. They have little alternative but to join earlier migrants in the growing number of slums or **informal settlements**. These settlements are built using whatever materials are available (such as corrugated iron, timber and plastic sheeting). They are not routinely provided with services, although at times a few basic services (such as stand pipes and public toilets) may be provided. However, these are usually insufficient in number for the size of the settlement.

The people living in these informal settlements have no legal right to occupy the land and can be forcibly removed by local authorities.

Above all else, it is the increase in the numbers of people living in such settlements that poses the greatest challenge to urban authorities. One third of all city dwellers in LEDCs live in informal settlements. At a global level, this amounts to 1 billion people, 55 million of whom were added since 2000. The rapid growth of such settlements has been described as 'the urbanisation of poverty'. As Figure 49 shows, there are distinct regional variations in the proportions of informal settlements.

Figure 49: Numbers of informal settlers as a percentage of urban population for selected regions

Source: Figures from 'World Cities Report 2016', UN Habitat, http://wcr.unhabitat.org

Area	Informal settlers (millions)	% of urban population living in informal settlements
Sub-Saharan Africa	200	56
Southern Asia	191	32
Eastern Asia	252	27
Latin America and Caribbean	105	21
South East Asia	84	29
Western Asia	37	25
Northern Africa	11	12

Exercise

1. Describe the regional variations in the percentage of urban population living in informal settlements shown in Figure 49.

While there is no doubt that conditions in informal settlements are extremely difficult, some observers have suggested they do provide some economic advantages. For one thing they offer the opportunity for rural migrants to access the opportunities of work and services which, although inadequate, are still better than in most rural areas. Secondly, they provide a pool of labour for new industries. Thirdly, for some, life in an informal settlement may only be a temporary hardship. Figure 50 shows three possible stages that new migrants may experience:

Figure 50: Changing priorities by residents of informal settlements

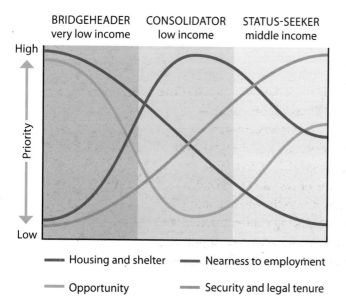

1. Bridgeheader: This is a new migrant to the city. The migrant's priorities include being close to opportunity, possibly in the centre of the city, and having some basic shelter.

2. Consolidator: Over time the new migrant earns some money and moves to an established informal settlement. The migrant's priorities now include having some security of residence, while still being close to work to avoid transport costs.

3. Status Seeker: The migrant now wants to be fully integrated into the urban way of life. The migrant has a well-paid job and the main priority is to obtain better quality housing.

It should be noted that not all residents of informal settlements will necessarily pass into the third stage.

Economic activity

Many migrants arrive in the cities with few of the necessary skills required for the limited number of jobs available and often find it difficult to gain full-time employment. Informal settlements have an important role in the economy of many LEDCs, providing a cheap and plentiful labour supply in the formal sector. The readily-available, low-cost labour force is an attraction for multi-national companies in some cities, but the pace of urbanisation is such that the demand for jobs rapidly outstrips the supply. The lack of employment opportunities in regulated or formal sector jobs has driven many to seek work in the informal sector, for example street-selling, shoe-shining, begging and prostitution. However, it has been argued that life and economic opportunity may still be better than in the rural areas. Opinion is divided between those who believe informal settlements are a 'ladder out of poverty' or a 'poverty trap' (Figure 51).

Figure 51: Contrasting perspectives on the role of the informal settlement

	Ladder out of poverty	Poverty trap
Migrants	Rural migrants come to the cities with ambition and the expectation of better prospects. Often, men are the first to move and other family members follow once he is established.	Unskilled migrants struggle to find a job in the formal sector. Working in the informal sector is unreliable, lowly paid and sometimes dangerous. Some are unsuccessful in finding a job and become destitute.
Settlements	The new migrants are prepared to endure the hardship of life in informal settlements for a short period of time.	The migrants are forced to live in informal settlements with no security of tenure and few amenities. Many of these are in unsafe areas, prone to landslides and flooding.
Economic activity	Migrants are prepared to accept any type of paid employment in the formal or informal sector.	Slum dwellers are often shunned by employers and become marginalised.

Service provision

Unemployment is only one problem caused by the pace of urbanisation in LEDCs. The large number of people moving into the cities puts added demands on essential services such as clean water supply, sewerage, waste disposal, health care and education. These services require money and expertise, which are not always readily available. One striking characteristic of LEDC cities is the growth of very large or mega cities. A study of city size in many LEDCs shows one or two very large cities, while the remainder are very much smaller. This is partly due to the uneven nature of development in the LEDC, which results in urban resources being concentrated in the larger cities. Once this process begins, these larger cities exert an even greater pull or attraction for migrants. The end result is to further increase the demand for jobs and other services in just a few cities.

Exercise

1. Study Figures 52 and 53 relating to employment rates and monthly income bands in urban and rural areas in South Africa.

   To what extent does the information provided in these figures support the idea that informal settlements are a 'ladder out of poverty'?

Figure 52:
Employment rates in urban (formal and informal) and rural areas 2008–2013

Source: Data from Quarterly Labour Force Survey, Statistics South Africa

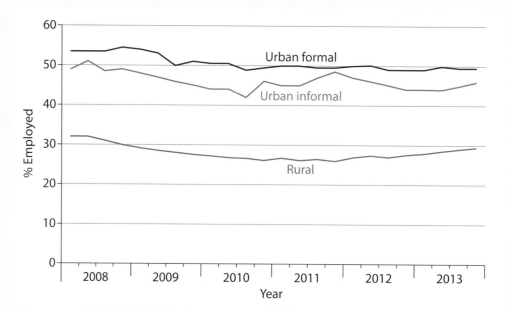

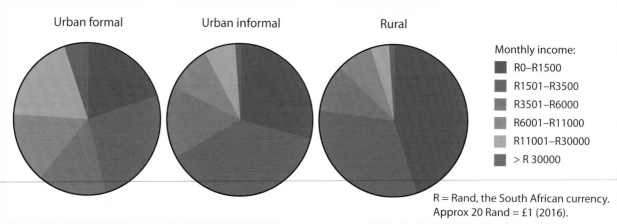

R = Rand, the South African currency.
Approx 20 Rand = £1 (2016).

Figure 53:
Monthly income bands of employed adults by type of residential area 2015

Source: Data from Quarterly Labour Force Survey, Statistics South Africa, 2015

2. Study Tables I and II.

   (a) Contrast the urbanisation process in the LEDCs with that in the MEDCs.

   (b) Explain why the situation in the LEDCs presents a challenge to the authorities in terms of service provision and economic activity.

Table I: Urbanisation in MEDCs

Year	% urban population	Numbers in cities
1750	10	15 million
1950	52	423 million

Table II: Urbanisation in LEDCs

Year	% urban population	Numbers in cities
1950	18	309 million
2030 (estimated)	56	3.9 billion

3. Study the information on Ezbet El Haganna, an informal settlement in Cairo, in Figure 54.

   (a) Describe the social and economic problems of the inhabitants of this informal settlement in Cairo.

   (b) Describe and explain the problems caused by rapid urbanisation in a LEDC city.

   (c) How do the issues in a MEDC city differ to those in a LEDC city?

Figure 54: Ezbet El Haganna – an informal settlement on the periphery of Cairo (Egypt)

Source: State of the World Population – report 2007

Ezbet El Haganna is one of the informal settlements on the north-eastern edge of Cairo, with a population of one million people. The following points illustrate the poverty of the inhabitants:

- 80% of households had no running water in their house.
- 60% were not connected to a sewerage system and raw sewage flowed in the streets.
- 52% of those aged nine and over were illiterate (more than twice the average figure for Cairo as a whole).
- 40% of the population was under 15.
- Average monthly household income was the equivalent of 44 US $ (the equivalent figure for Cairo was 206 US $).
- Population density exceeded 220,000 per km^2.
- 30% of families lived in one single room, sharing toilets.
- 65% of those employed were working in the informal sector.
- Many residents regularly suffered frequent health problems (especially diarrhoea, kidney, liver, eye and skin ailments). Infant and general mortality rates were thought to be high although no accurate figure is available, as births and deaths were often not recorded.
- Many dwellings were in need of structural repair. Between 3,000–4,000 dwellings had no roofs.
- Only 1,200 of the residents had voting cards – a requirement to vote at elections.
- Most residents, especially women, had no legal documents, such as birth certificates or identity cards.

References

Geofile article:

'Issues Facing World Cities in the Fastest Growing LICs', *Geofile* 705, series 2, 2013–2014

Additional information on urbanisation in LEDCs:

www.unpopulation.org
http://www.econ3x3.org/article/informal-settlements-poverty-traps-or-ladders-work
https://sustainabledevelopment.un.org/content/documents/745habitat.pdf

World Urbanization Prospects, Population Division, Department of Economic and Social Affairs, United Nations:

http://esa.un.org/unpd/wup/index.html/
http://unhabitat.org/urban-themes/housing-slum-upgrading/

The following websites provide short articles on informal settlements:

http://wcr.unhabitat.org/main-report/ 2016
www.globalissues.org/news/2015/09/03/21453
www.globalissues.org/news/2015/05/22/21024
www.globalissues.org/news/2015/05/05/20953

CASE STUDY: Issues and Challenges in a LEDC, Mumbai

Mumbai, formerly Bombay, is the commercial capital of India and a global financial centre. About one third of the tax revenue and 40% of the foreign trade of India as a whole are generated in Mumbai. Its location on the west coast of India facilitated the city's growth as a major trading port and Bombay (the city did not change its name until 1995) became an important cotton exporting city during British Colonial rule. The opening of the Suez Canal in the late nineteenth century further enhanced the city's trading potential and Bombay became the world's most important cotton exporting centre. The textile mills are mostly closed now and Mumbai has become a modern city with modern industries and services including IT, financial services, engineering, media and scientific research. Mumbai is also the headquarters of India's growing film industry, known as Bollywood.

India has experienced considerable economic development in the last 20 years and is now one of several emerging market countries among LEDCs known as BRICS (see page 251). This economic development is uneven and many parts of India, especially rural areas, remain largely unaffected by this recent growth in prosperity. The opportunities for a better lifestyle, including a reliable source of income, access to education and health services, combined with better infrastructure in Mumbai, have encouraged large numbers of rural migrants to move to the city. These migrants are generally in the economically active age groups, leading to high rates of natural increase. Natural increase and migration have both contributed to the rapid increase in Mumbai's population, which doubled between 1991–2013. Between 1991–2001, rural urban migration accounted for 43% of the city's population growth. In 2016, the Mumbai metropolitan area had a population of 22 million and it is predicted to reach 26 million by 2025, making it the second largest urban area in the world.

Informal settlements in Mumbai

People migrate to Mumbai to escape rural poverty and seek better opportunities but cities concentrate the poor in deprived areas, with limited opportunities for social mobility. When the migrants arrive in Mumbai, the reality of their situation is often well below their expectations. Faced with a large influx of migrants, Mumbai, like many LEDC cities, has struggled to provide adequate housing, services and employment. Surrounded on three sides by water, Mumbai has limited building space and house prices are very high as a result. It is a city of great contrasts between those who can afford to pay the very high house prices and those who have to live in the many crowded informal settlements, which have grown rapidly in and around the city. Informal settlement dwellers account for over 40% of Mumbai's population (9 million people) living in 2,000 settlements. These 9 million people reside in just 6% of the land area of the city, leading to very high population densities. Mumbai has one of the highest proportions of informal settlement dwellers in LEDCs. This is largely the result of the scarcity of available building land and affordable housing in the city.

Squatter settlements have been growing in Mumbai since the 1940s. The squatters built their shelters on land unsuitable for development, such as marshy areas, close to railway lines or hillsides. Whilst they are found across Mumbai, over 80% of squatters live in the suburbs of Mumbai. With no legal rights these informal settlements were frequently cleared by the authorities. This was not an effective way of dealing with such settlements, as the displaced people usually rebuilt their dwellings elsewhere in the city. In 1976, the government issued an identification number to each squatter household. This effectively legalised the squatter settlements, which were now known as slums. The largest (most populated) and best-known slum is Dharavi.

Dharavi, located close to the centre of Mumbai, between the two main railway lines, began as a small settlement set aside for Indian workers during British rule. It was close to the tannery and pottery works, both of which were the source of considerable amounts of air and land pollution. With the growth of modern Mumbai, Dharavi is now fully surrounded by commercial buildings. Land that was formerly undesirable has become highly valued and as such the future of Dharavi and its inhabitants is under threat. There is no official figure for the population of Dharavi but most estimates suggest that close to 1 million people live there, making it Asia's second largest informal settlement. The total area of Dharavi is 200 hectares and with 1 million inhabitants, this results in a population density of 5,000 per hectare.

Figure 55: Map showing the informal settlements in Mumbai

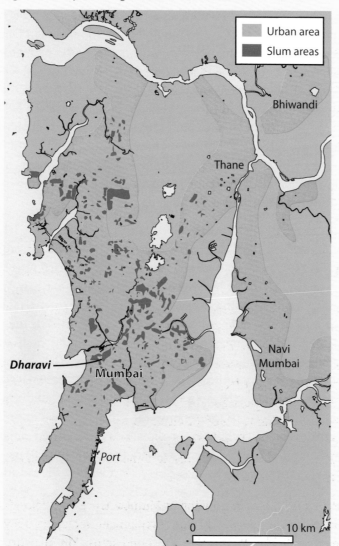

Figure 56: Dharvavi,
showing open drains

Source: © A Savin,
Wikimedia Commons

Service provision

Within Dharavi, service provision is extremely limited and the settlement is still
growing as more people move into the area. There is considerable overcrowding.
Houses are typically 10–20 m². One account of living conditions in Dharavi
describes 12 people living in 8 m² (for comparison, this is approximately half the
area of a small, single, domestic garage). Rooms are frequently used for several
purposes, such as eating, sleeping and working. Much of the housing is permanent
but poorly constructed. Some housing is built on unsafe areas such as hillsides,
marshy areas or dangerously close to power lines. Others obstruct natural drainage
systems, cause traffic disruption and present fire hazards. The roofs of the houses
often use asbestos, which can damage the human respiratory system. The electricity
supply does not comply with the required safety standards, resulting in a fire risk,
especially in the high density, wooden housing.

Inadequate sanitation is a major issue. Over 73% of the residents of Dharavi
depend on public toilets. Only 14% of these have piped water. It is claimed that
16 public latrines serve over 3,000 people. There is no organised waste collection
and residents dump their waste into the open sewers. Toxic waste from tanneries
and other industries, along with human waste flow into open sewers. Some of this
waste seeps through cracked pipes into the water supply, which is then used for
drinking, washing and cooking. During the wet season of the monsoon, Dharavi is
prone to flooding and this makes the situation even worse. The water supply comes
from standpipes and water is only available for a few hours each day. One third of
the population have no access to tap water. These unsanitary conditions provide
ideal breeding grounds for mosquitoes, rats and other vermin. Outbreaks of
malaria, hepatitis, cholera, typhoid and polio are common, and related to
inadequate sanitation. A drug resistant strain of tuberculosis (TB) has also been
diagnosed. Disease spreads rapidly in the overcrowded living areas and, with
limited access to medical care, life expectancy is reduced. Life expectancy in
Dharavi is 50 years compared to 67 years for India nationally. Child mortality is
twice the rate of Mumbai as a whole. In a settlement of 1 million people, there are
no hospitals and the residents have to rely on the limited availability of free health
care outside of Dharavi.

Informal settlers generally have low levels of literacy, limiting their opportunities to escape the 'poverty trap'. In Dharavi, less than half of the children attend primary school and only one third of the population complete 10 years of schooling. Literacy levels at 60% are 14% lower than the national average for India. Mobile schools, funded partly by Non-Governmental Organisations (NGOs), have been introduced with the aim of providing basic education to all children in such areas. However, the drop-out rate is high, as children are still seen as a labour supply and income for their families.

Economic activity

Employment opportunities are very limited in the informal settlements. In Dharavi, many are employed in the settlement's vibrant informal sector, which involves the manufacture of leather goods, pottery, sewing cotton and recycling. There are over 5,000 businesses operating in 15,000 single room premises. Recycling is a major industry in Dharavi and child labour is used to collect and sort rubbish from the city dumps. This is dangerous and unsanitary work, containing hospital and industrial chemical waste. Once sorted, the waste is recycled, often involving the melting of plastics which give off toxic fumes. Informal sector work is unregulated and unreliable. There are no health and safety guidelines and hazardous waste is disposed into the open sewers. Dharavi has been successful in exporting its manufactured goods, generating millions of dollars but much of this money is unevenly distributed, and the lives of the poorest have not improved, with over 40% of the residents living below the poverty line.

Figure 57: Aerial view of Dharavi showing cramped conditions. The commercial district is visible in the background.

Source: YGL Voices

A significant number find employment outside of Dharavi. Men traditionally undertake low paid and undesirable employment, for example, as drivers, labourers and guards. Women frequently find employment as cleaners and shop workers in the affluent parts of the city. Some women carry building materials on building sites.

There have been many attempts to deal with the problems of informal settlements in Mumbai. Early attempts included the compulsory clearance of some settlements but, as stated earlier, this was deemed unsuccessful as the displaced people rebuilt their houses elsewhere in the area. The provision of basic sanitation and cheap housing was seen as a more positive approach but the demand far exceeded the supply. Currently, there are major plans for the Dharavi settlement, partly because it is built on a valuable site, close to the new business centre of Mumbai. A number of plans have been considered but each has proved to be controversial. There is a conflict of interest between the developers, who would prefer to resettle most of the residents in high rise apartment blocks in a small part of this highly valued site, and the residents, who wish to retain control of all of Dharavi but with improved amenities and living conditions.

Figure 58: Pottery works in Dharavi
Source: M M

Exam Questions

1. With reference to your urban case study from a LEDC, describe and explain the main issues and challenges found in a LEDC city. [15]

2. With reference to your urban case studies, describe how the issues and challenges found in a LEDC city contrast with those in a MEDC city. [15]

References

Geofile articles:

'Mumbai: Case study of a Megacity', *Geofile* 696, series 32, September 2013

'Slums in Mumbai' in 'Urban slums reports: the case of Mumbai, India':
http://www.ucl.ac.uk/dpu-projects/Global_Report/pdfs/Mumbai.pdf

The Challenge of Slums – Global Report on Human Settlements:
http://unhabitat.org/wp-content/uploads/2003/07/GRHS_2003_Chapter_01_Revised_2010.pdf

Short accounts of relevant topics on Dharavi:

http://news.bbc.co.uk/1/shared/spl/hi/world/06/dharavi_slum/html/dharavi_slum_intro.stm

https://www.theguardian.com/cities/2015/feb/18/best-ideas-redevelop-dharavi-slum-developers-india

https://www.theguardian.com/global-development-professionals-network/2015/oct/30/health-in-indian-slums-inside-mumbais-busiest-public-hospital

https://www.theguardian.com/cities/2014/apr/01/urbanist-guide-to-dharavi-mumbai

https://www.theguardian.com/cities/2014/nov/28/slum-loaded-term-homegrown-neighbourhoods-mumbai-dharavi

Country	GNIpc US$	Rank	GNI (PPP) pc US$	Rank
Norway	103,050	4	65,970	10
Switzerland	90,670	6	59,600	12
Kuwait	52,000	18	87,700	5
Canada	51,690	19	43,400	28
UK	42,690	31	38,370	37
Bangladesh	1,080	181	3,330	169
Tanzania	930	187	2,530	181
Ethiopia	550	203	1,490	201
The Gambia	440	206	1,560	198
Malawi	250	213	790	209

Figure 63:
Comparisons between
GNI per capita and GNI
(PPP) per capita 2015

Source: Data from
'Table 1.1 World
Development Indicators:
Size of the economy',
http://data.worldbank.org,
September 2015

Percentage of the population living on $1.90 a day (poverty threshold 2015)

One of the Millennium Development Goals was to reduce the proportion of the world population living in poverty (Figure 64). In 2000, the threshold for poverty was set at $1 per day but was later revised to $1.25 and $1.90 in October 2015. This measure defines the percentage of the population living on less than $1.90 a day. It gives an overall picture of levels of poverty in different countries and it can be used to show trends over time. However, it gives no indication of the distribution of poverty within a country and it shares the same limitations as the other economic measures. In addition, obtaining accurate and reliable data is often difficult in the most deprived regions.

Region	1990	2002	2012
East Asia and Pacific	60.6	29.2	7.2
Europe and Central Asia	1.9	6.2	2.1
Latin America and the Caribbean	17.8	13.2	5.6
North East and North Africa	6.0	No data	No data
Sub-Saharan Africa	56.8	57.1	42.7
East Asia	50.6	40.8	18.8
World	37.1	26.3	12.7

Figure 64: Percentage
of the population
living on $1.90 a day
(poverty threshold
2015)

Source: Data from
'Table 2.8 World
Development
Indicators: Poverty
rates at international
poverty lines Part 2',
http://data.worldbank.org

Social measures

Social measures are increasingly used as indicators of development because they show the impact of development on society.

Life expectancy at birth

Life expectancy at birth is one of the most common social measures used. It refers to the number of years a person is expected to live if the social conditions at the time of birth remain constant throughout a person's life. It is a clear reflection of the social conditions existing at a given time. The figures for life expectancy are traditionally higher in MEDCs than in LEDCs. However, there has been a steady increase in life expectancy worldwide. In 2015 the global figure was 71 years. Some of the largest increases have occurred in Sub-Saharan Africa, where the figure is now 57 years. This region suffered a fall in life expectancy from 1980–2005 but since then life expectancy figures have increased, as a result of increased availability to medical care and AIDS awareness. Gender differences in life expectancy are also diminishing in LEDCs.

Life expectancy is a very reliable measure, especially for LEDCs, where changing trends are evident. Figure 65 gives figures for selected countries for 1980, 2000 and 2014. The decline in AIDS related deaths is reflected in the increased life expectancy calculation for countries in southern Africa. The Russian Federation had experienced declining socio-economic conditions in the 1990s and this was reflected in the falling figures for those countries from 1980–2000. Improvements in living conditions in Russia are now reflected in the current higher figure. However, it is still an average figure and may well hide variations in society. Since it is calculated at birth there is no accounting for new developments in medicine that may occur. As conditions in LEDCs improve, a vaccination programme could eliminate some diseases or have a dramatic impact on survival rates from endemic disease. Life expectancy figures obviously cannot take account of this.

Figure 65: Life Expectancy figures for selected countries 1980, 2000 and 2014

Source: 1980–2000 data from 'Life expectancy at birth', UN Development Programme, Human Development Reports, http://hdr.undp.org/en, 2014 data from '2015 World Population Data Sheet', Population Reference Bureau, www.prb.org

Country	1980	2000	2014
Norway	76	79	82
Ireland	72	77	81
Estonia	69	70	77
Saudi Arabia	63	72	74
Russian Federation	67	65	71
Mexico	66	74	75
Botswana	61	49	64
South Africa	57	56	61
Zambia	51	42	53
Mali	40	49	53
India	55	62	74

Progress in education is seen as critical for human development. There are various measures used to chart a country's progress in this field:

- **Enrolment in primary education** – The percentage of children enrolled in primary school.
- **Adult literacy rates** – The percentage aged over 15 who are literate.
- **Youth literacy** – The percentage under 15 who are literate.

Once again, progress appears to be made in most countries whichever indicator is used. Much of this progress is due to the work carried out during the Millennium Development Goals (see pages 229–230). However, women still fall behind men in many countries. In Afghanistan female adult literacy was 18% in 2011 compared to 32% for men. Globally, in 2012, 80% of women aged 15 and over had basic literacy compared to 90% for men. Figures 66–70 show patterns and trends in these social indicators.

Figure 66: Adult literacy rates

Source: Data from 2013 UN Human Development Report

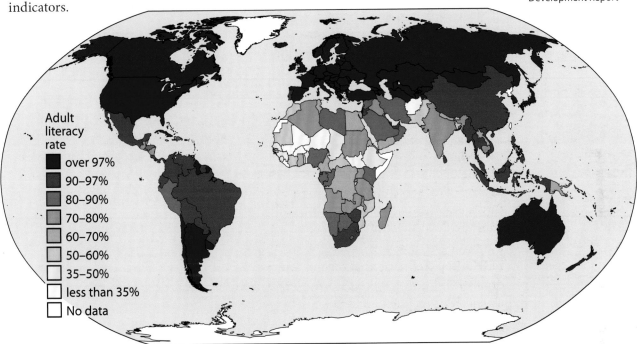

Exercise

1. Study Figure 67 which shows enrolment rate in primary school education 1990–2015.

   (a) Describe the changes from 1990–2015.

   (b) How effective is this indicator in demonstrating levels of development across major world regions?

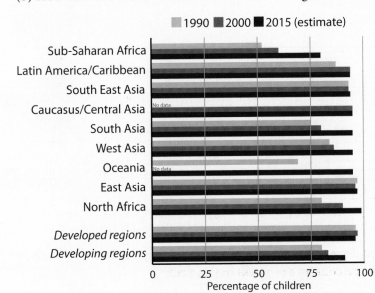

Figure 67: Enrolment rate in primary school education 1990, 2000 and 2015 (percentage)

Source: Data from UN Millennium Development Goals Report 2015 (no data available for Oceania 2000)

2. Study Figures 68, 69 and 70 relating to literacy levels by region and gender.

(a) Describe the pattern of adult literacy in 2011, as shown in Figure 68.

(b) Using named examples from Figures 69 and 70, describe the trends in male and female adult literacy between 1990 and 2015.

(c) Using examples from these resources, evaluate the effectiveness of adult literacy as a measure of development.

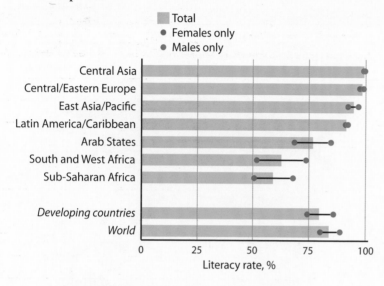

Figure 68: Adult literacy rates by region and gender 2011

Source: Data from UNESCO Institute for Statistics, May 2013

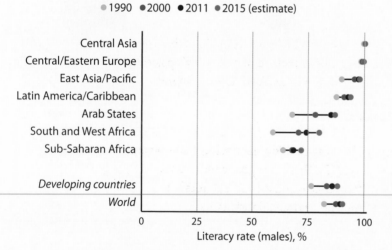

Figure 69: Adult literacy rate of the male population by region, 1990–2015 (estimated)

Source: Data from UNESCO Institute for Statistics, May 2013

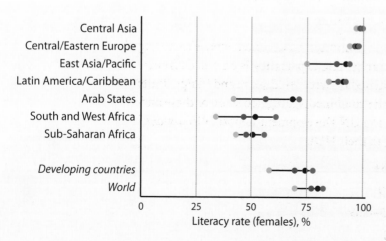

Figure 70: Adult literacy rate of the female population by region, 1990–2015 (estimated)

Source: Data from UNESCO Institute for Statistics, May 2013

3B Reducing the development gap

Students should be able to:

(i)   understand the aims of the Millennium Development Goals and evaluate the impact of any two of these as a means of improving global development

(ii)   understand the aims of the 2030 Agenda for Sustainable Development and explain how the Global Goals build on the Millennium Development Goals

(iii)   explain the different roles that globalisation and aid can have in influencing development in LEDCs

1. The Millennium Development Goals (MDGs)

The year 2000 was an important landmark in human history. The new Millennium was welcomed globally by much celebration and optimism for a new future. The world had recovered from the economic recession of the 1980s. The collapse of the Berlin Wall in 1989 marked the beginning of the end of communism in eastern Europe. Germany was reunited and many former Soviet Union countries won their right to independence.

For many in LEDCs the new millennium brought little cause for optimism. In 1999 1.75 billion people, or half of all of the population in LEDCs, were living on less than $1.25 dollars a day, only slightly fewer than in 1990. Infant mortality rates and maternal mortality rates remained stubbornly high. Education was not universally available and the role of women displayed serious inequalities. The environmental impacts of global warming threatened many fragile environments, especially those in developing countries that lacked the economic or technological resources to deal with such issues.

Many LEDCs were heavily indebted to international banks caused by large-scale borrowing to finance development projects in the 1960s. Higher interest rates and increased oil prices added to the debt problems. Developing nations also had trade issues. In the west, the cost of manufactured products increased but the price of primary goods or commodities fell. As LEDCs exported mostly commodities and imported manufactured goods, their balance of trade deteriorated, pushing them further into debt. Some countries were so indebted that repaying the debt seemed impossible. Any hopes of social development, such as education and health care, were often sacrificed to pay off interest on loans.

A number of initiatives were introduced to ease the debt burden of some LEDCs but it became apparent that a more coordinated approach was needed to steer development in LEDCs. This coordinated approach would provide both guidance and financial aid.

On September 2000, world leaders at the UN launched **the Millennium Development Goals (MDGs)** (Figure 75). These consisted of eight goals, with each goal sub-divided into a number of targets. Progress on each target was monitored by a series of indicators.

Figure 75: The Millennium Development Goals

Source: From 'Millennium Development Goals', by Millennium Project, UNDP Brazil, © 2002 United Nations. Reprinted with the permission of the United Nations.

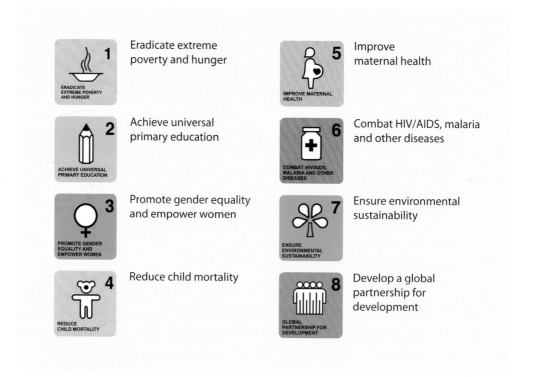

1. Eradicate extreme poverty and hunger
2. Achieve universal primary education
3. Promote gender equality and empower women
4. Reduce child mortality
5. Improve maternal health
6. Combat HIV/AIDS, malaria and other diseases
7. Ensure environmental sustainability
8. Develop a global partnership for development

Aims of the Millennium Development Goals

The MDGs had four main aims for human development in LEDCs:

- To reduce poverty and its associated problems, including lack of education, unemployment, hunger and disease. (Goals 1, 2, 4, 5 and 6)
- To promote gender equality. (Goal 3)
- To work towards environmental sustainability. (Goal 7)
- To establish a global partnership for development. (Goal 8)

The MDGs were to be financed through aid donated by the MEDCs to the World Bank and administered by the Development Assistance Committee (DAC). Financial support would also come from Non-Governmental Organisations (NGOs), such as Oxfam, and private donations, such as The Bill Gates Foundation. Debt relief programmes were to be extended in the hope that developing countries could contribute, in part at least, towards their own development. The MDGs attracted much publicity, particularly Goal 1, with campaigns such as 'Make poverty history'. World leaders met in 2005 to assess the progress of the programme, which was slower than expected at that stage. It was decided that a much greater level of funding was required if the goals were to be achieved by 2015.

The Millennium Development Goals differed from all other development initiatives in two ways:

- The 8 goals worked within the context of the UN definition of development, which sees economic development as a means to provide social development.
- The goals were to be achieved within a period of 15 years.

The impacts of the Millennium Development Goals on global development

Goal 1: Eradicate extreme poverty and hunger

Progress on this goal was measured against the following targets:

1. **(a)** Halve, between 1990 and 2015, the proportion of people whose income is less than $1 a day.

   (b) Achieve full and productive employment and decent work for all, including women and young people.

   (c) Halve, between 1990 and 2015, the proportion who suffer from hunger.

Target 1 (a): Halve, between 1990 and 2015, the proportion of people whose income is less than $1 (adjusted to $1.25) a day

Significant improvements have been made with regard to the eradication of extreme poverty, which is defined in the MDGs as living on less than $1 a day. In 1990, 1.9 billion people fell below this poverty line: the target to reduce this number by half was achieved by 2010, five years ahead of schedule. By 2015, the number living in extreme poverty had fallen by more than 1 billion from 1990 to 836 million. On a world scale there has been a 57% decrease in extreme poverty but some regions have shown very dramatic falls in poverty levels. South East Asia, the first developing region to reach the target, has seen an 84% reduction in extreme poverty levels since 1990 and in China the reduction was 94%.

However, there are important variations in the level of success. For example, Sub-Saharan Africa has not been able to achieve the target reduction, despite some improvement. In 2015, 41% of the region's population were deemed still living in extreme poverty. In addition, the indicators used to gauge the levels of success of Target 1 (a) do not take into account the regional variations that may occur within a country. This is particularly evident in countries such as China, where there are marked regional variations in wealth. Many LEDCs have large numbers of people living in informal settlements in urban areas and in vulnerable situations in rural areas. Progress has also been affected by a number of conflict situations that have created the largest refugee crisis in modern times. The United Nations High Commission for Refugees (UNHCR) estimated that conflicts such as those in Afghanistan, Syria and Somalia had resulted in more than 60 million refugees by the end of 2015. Poverty levels still show a marked gender imbalance. It is estimated that women are more likely to live in extreme poverty in 45 of 75 countries in LEDCs. In Latin America and the Caribbean, the ratio of women to men in poverty has increased significantly since 2000 despite the overall rate of poverty decline in that entire region.

Note:

The specification requires the evaluation of the impacts of any **two** Millennium Development Goals on global development. Goals 1 and 3 are discussed here.

Figure 76: Proportion of people living on less that $1.25 a day, 1990, 2011 and 2015 (percentage)

Source: Data from Millennium Development Goals Report 2015

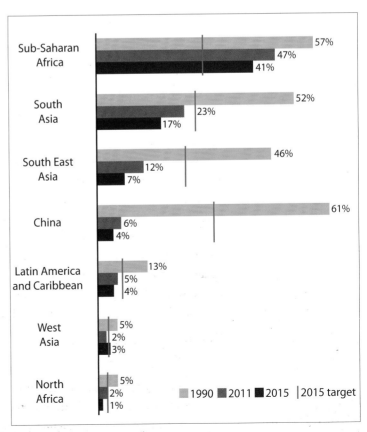

Target 1 (b): Achieve full and productive employment and decent work for all, including women and young people.

This was an ambitious target, even before the economic downturn of 2008, and not surprisingly it has not been achieved. Worldwide, there has been an increase in unemployment but the greatest increases have occurred in LEDCs, particularly South and East Asia. There are two main reasons for this:

- the growth in the economically active age groups in LEDCs.
- global economic problems have resulted in fewer new jobs.

Nevertheless, there have been some achievements associated with this target. Those in employment in LEDCs have seen an improvement in their situation. Wages are higher and there has been a two thirds decline in the number of workers living in extreme poverty in LEDCs and nearly half of the workforce now earns more than $4 a day. Progress across LEDCs has been uneven and many workers in Sub-Saharan Africa still remain close to the poverty level. Unemployment rates for women and youths remained stubbornly high across all developing regions.

Target 1 (c): Halve, between 1990 and 2015, the proportion of people who suffer from hunger.

On a world scale, this target is close to being achieved with the proportion of undernourished people, a widely used indicator for hunger, having fallen by almost half since 1990. However, approximately 780 million people in LEDCs are still suffering from hunger. Progress has slowed down since 2008 due to a combination of factors, including higher food prices, economic recession, extreme weather events, natural disasters and wars. Countries in East and South East Asia, along with North Africa and Latin America have reached this target but Sub-Saharan Africa, the Caribbean and parts of South Asia remain below the target.

Figure 77: Percentage of children under 5 years old with malnutrition 1990–2013

Source: Data from UNICEF, WHO and World Bank 2014

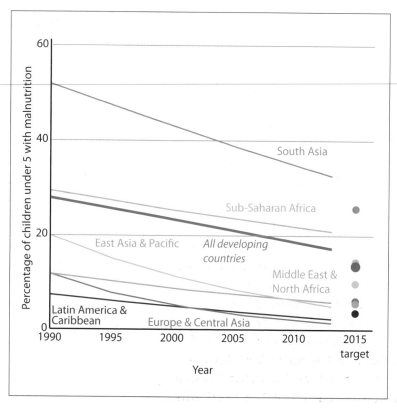

The proportion of underweight children, another indicator for hunger, has been reduced by almost 50% worldwide since 1990, close to MDG 1's target. However, due to population growth, the actual numbers of those underweight in some regions are actually increasing. Half of the world's underweight children live in South Asia and one third live in Sub-Saharan Africa. These regions also have the highest percentage of children suffering from malnutrition (Figure 77).

As stated earlier, this Goal attracted much media coverage and was supported by many NGOs. Considerable improvements have occurred in all LEDCs, even in those areas where the target has not been achieved, but there is still much to be achieved. The Global Goals aim to eradicate poverty and hunger everywhere by 2030.

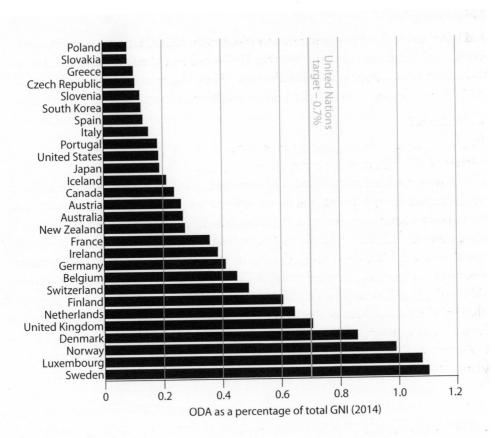

Figure 88: Aid as a percentage of GNI 2014

Source: ODA preliminary data 2015, OECD

2. Study Figure 89 relating to the percentage of total Official Development Assistance (ODA) donated to Sub-Saharan Africa and Asia 1995–2009.

How did the MDG programme from 2000 affect the percentage aid donated to these two regions?

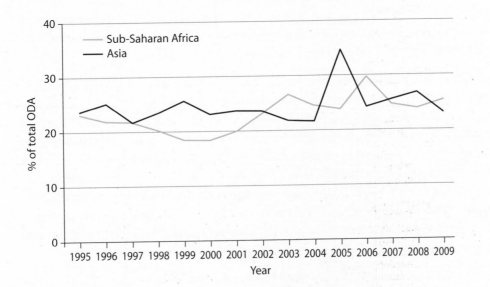

Figure 89: Official Development Assistance to Asia and Sub-Saharan Africa as a percentage of total ODA

Source: OECD-DAC 2010

2. Voluntary Aid

Voluntary aid comes from charitable organisations such as Oxfam, Trócaire and Action Aid. These are referred to as **Non-Governmental Organisations (NGOs)** and they rely entirely on voluntary contributions. These organisations provide emergency aid following a natural disaster, as well as long-term aid packages to provide clean water supply or irrigation schemes. Voluntary aid is a much smaller amount than official aid.

Impacts of aid in LEDCs

Both MEDCs and LEDCs want to contribute towards official aid programmes, partly for humanitarian reasons but also because their own countries will benefit from any economic improvements in the LEDCs through increased markets for their goods. There is evidence to show that aid packages, from whatever source, bring economic improvement to LEDCs. Some of the most dramatic examples are seen following a natural disaster, such as Typhoon Haiyan in the Philippines in 2014 and the Nepal earthquake in 2015. This type of aid provides emergency food and medical supplies in the immediate aftermath of the disaster but also assists the reconstruction of the region in the long term. Aid packages have also contributed to vaccination programmes, which have resulted in a decrease in infant mortality. The very significant progress made in many LEDCs during the MDG programme, in terms of poverty reduction and enrolment in primary school, is evidence of the importance of aid (see the MDGs on page 226). Official aid, in the form of money, incurs interest but at much lower rates than normal loans. In this way, official aid allows countries to borrow money without getting into serious debt and some countries have used aid for large-scale development projects. Aid works best in countries where there is effective democratic government and political stability. Unfortunately, in many LEDCs these are frequently absent.

There has been considerable criticism of aid as a means of encouraging development in LEDCs. Too often much of the financial aid is wasted, either through corruption, bad management, poorly thought-out schemes or spent on arms deals. One of the best-documented examples is that of former Zaire (Democratic Republic of Congo), where millions of pounds of aid money was stolen by the corrupt ruler in the 1970s and 1980s and never reached the poor for whom it was intended.

Until recently, much bilateral aid was 'tied aid', meaning that the aid had conditions placed upon it. It is estimated that tied aid accounted for about 40% of all aid donations to LEDCs in the 1990s. This usually meant that the recipient country had to buy manufactured goods from the donor country, or the recipient may have had to agree to terms of trade beneficial to the donor. Britain and other countries have been accused of linking aid packages with arms sales in LEDCs. In addition, multilateral aid from organisations such as the World Bank was often criticised for only providing aid to those countries which supported 'western style' politics. Many large-scale projects funded by the World Bank in the past have had serious environmental repercussions. Others claim that aid creates a dependency culture and prevents initiative.

In response to these criticisms, aid and its administration have changed significantly since 2000. Aid packages are no longer a simple transfer of money to a LEDC but are linked to development targets. Greater attempts have been made to develop a partnership between the aid donor and the recipient country. Much of this has stemmed from the MDG programme and was part of a strategy to make aid more effective. Since 2002, much work has also been done to make aid donations more transparent and to reduce the amount of 'tied aid'. Environmental issues were an important concern in the MDG programme. Agenda 2030, UN's action plan for development from 2015–2030, focuses on sustainable development through The Global Goals and any new projects will have to conform to the ideals of sustainability.

Exercise

Review the information given at each of the following websites:

Disasters Emergency Committee* – http://www.dec.org.uk/

Oxfam UK and Oxfam Ireland – http://www.oxfam.org.uk/ and https://www.oxfamireland.org

WaterAid UK – http://www.wateraid.org/uk

The Disasters Emergency Committee (DEC) is an umbrella organisation in the UK that works on behalf of 13 charitable organisations.

1. What type of aid is provided by the above organisations?

2. Review the section on the Millennium Development Goals (pages 226–230). Use this information, along with examples from the above websites, to discuss how aid programmes can help LEDCs. You must make reference to places in your answer.

References

Youtube videos on Bhopal:

https://www.youtube.com/watch?v=UH5LPwdVnqI

https://www.youtube.com/watch?v=UvhPqmFkhes

Interactive graphs showing the amount of aid donated by selected countries:

http://www.compareyourcountry.org/oda?cr=oecd&lg=en

Geofile articles:

'International Agencies: How Poverty is Addressed on a Global Scale', *Geofile* 681, series 31, 2012–2013

'Poverty and Health – The Impact of Inequality', *Geofile* 689, series 31, 2012–2013

'Gender Equality and International Development', *Geofile* 660, series 30, 2011–2012

CASE STUDY: National LEDC case study: Tanzania

Tanzania belongs to the group of counties collectively referred to as Sub-Saharan Africa. The country is bordered to the north by Kenya and Uganda, to the west by The Democratic Republic of Congo, Rwanda and Burundi, and to the south by Mozambique, Malawi and Zambia. Most of the country is mountainous, the highest land occurring in the north east including Africa's highest peak, Mount Kilimanjaro. The East African Rift Valley occupies a large part of western Tanzania. The centre of the country is largely highland plains. Tanzania has three of Africa's largest lakes; Lake Victoria, Lake Tanganyika and Lake Nyasa. The only areas of lowland are found on the east coast. Climate is tropical throughout but with considerable regional variations due to relief and latitude.

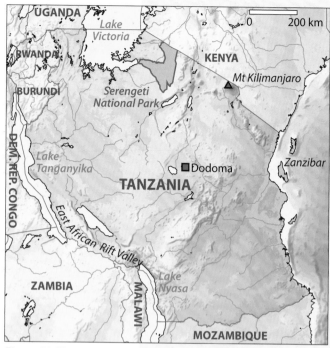

Figure 90: Tanzania and surrounding countries

Source of relief map:
© Sémhur / Wikimedia Commons / CC-BY-SA-3.0

Tanzania, formerly known as Tanganyika, was colonised by Germany until 1919 and then by Britain. The country gained independence in 1961 and became known as Tanzania. The capital city is Dodoma.

Tanzania is a poor country, belonging to the low income group of countries with some valuable natural resources including iron ore and coal, as well as reserves of gold and gemstones including tanzanite and diamond. Natural gas has recently been discovered off-shore in the south of the country. However, all of these are currently underdeveloped. Agriculture is the mainstay of the economy and agricultural products account for 85% of exports. Tourism is a major earner of income mostly associated with wildlife safari holidays. Almost 40% of the land is protected for conservation including the Serengeti National Park.

Figure 91: Selected Development indicators for Tanzania

Source: UN Human Development Report 2014 and Population Data Sheet 2015, Population Reference Bureau

GNI per capita $ US	Human development index	Life expectancy (years)	Infant mortality rate per 1000	Maternal mortality rate per 100,000	Percentage of population in poverty	Total fertility rate	Enrolment in primary education %
930	.448	62	35	410	28.2	5.2	90

The role of globalisation in Tanzania

LEDCs such as Tanzania require assistance if they are to progress in economic development. Tanzania has not experienced much political instability or ethnic issues that are common in many African countries. This has enabled the country to attract outside investment and several MNCs. Globalisation, a process where the world's economies are linked on a global scale, has had a number of impacts on Tanzania's development. Several countries including the UK, USA and China have made large investments in development projects. These projects, financed by private companies, include development of port facilities, mining projects, agribusiness and communications networks. Chinese firms invested $4 billion dollars in Tanzania in 2014. Whilst these developments have many positive benefits, some groups have raised concerns over their long term impacts.

Two major port developments (Figure 92) at Dar es Salaam (Figure 93) and Bagamoyo, financed by China, have proved to be controversial. Dar es Salaam is seen as a key entry point to East Africa but the original port could not handle large modern shipping. The Chinese investment of $400 million will greatly increase the capacity of the port and will enhance trade between China and Tanzania. In addition, Chinese investment has been used to construct a railway

populous country and therefore has a large potential market. Nigeria has considerable political importance and is predicted to become a major trading hub in Africa.

The MINT countries have all shown dramatic improvements in development in recent years. Indonesia and Mexico have modernised and implemented important economic reforms that are predicted to enhance their manufacturing industries. Indonesia also has large mineral supplies and a readily available market in China. Nigeria has considerable oil supplies, which should help to promote industrial development. Turkey's development stems from a growing domestic market and a booming construction industry.

Total GDP ($US) 2012

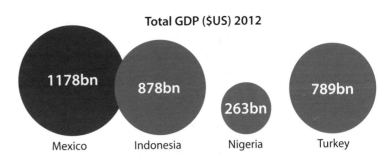

Total population (millions) and population growth rate (%) 2013

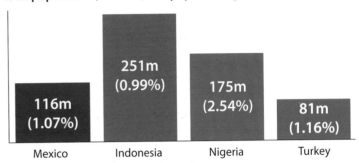

Figure 99: Selected indicators for MINT countries

Source: Data from World Bank and CIA World Factbook

Some observers have expressed doubt over the longer term future of these emerging markets. Economic development has come at a price. China in particular has based development on the use of fossil fuels and many of the major cities have serious air pollution problems. There are concerns over those countries that rely heavily on raw materials or energy supplies. The fall in oil prices in 2015 will have had a negative impact on the countries relying on oil revenues. This highlights the need for a greater level of diversification in industry to offset such fluctuations in the price of raw materials.

Recent global economic problems are having an impact on many of the emerging market economies. China and Brazil have been significantly affected and the long-term impacts are not yet fully analysed. The indicators used to show economic growth in these countries often hide great disparities in income distribution. China displays marked variations in wealth between urban and rural areas. In addition, many emerging market economies have poor track records for human rights. Economic success in these countries has brought prosperity to millions in LEDCs but the long-term sustainability of their progress has yet to be tested.

Exercise

1. Define each of the following terms:
   - emerging markets
   - BRICS and MINT countries

2. Discuss the extent to which you agree with the following statement:

   "Emerging markets in LEDCs have experienced an increase in prosperity but this may not be sustainable."

References

A number of short articles on Emerging Markets are available from the following websites:

http://www.theguardian.com/business/emerging-markets

https://www.globalsuccess-club.net/en_GB/mint-countries

http://www.bbc.co.uk/news/magazine-25548060

http://www.theguardian.com/business/2014/jan/09/mint-condition-countries-tipped-economic-powerhouses

CASE STUDY: National study of a BRICS country, China

Between 1949 and 1976, China was led by Chairman Mao, who followed a strict form of communism and the economy stagnated. Two years after Chairman Mao's death, Deng Xiaoping took leadership and, while still adhering to communism, began a series of reforms that would transform China into the most important of the emerging market economies and ultimately into the world's second largest economy by 2015.

Reasons for China's economic growth

China's economic growth is largely due to two main reasons:

1. Economic reform

Under the leadership of Deng Xiaoping (1978–1992) a number of important economic reforms were introduced. In the late 1970s China established a number of **Special Economic Zones (SEZ)**. Initially, there were five SEZs but later they were extended along much of the south east coast. Each was provided with essential infrastructure to attract industrial development. These areas, including Shanghai, adopted economic policies which encouraged foreign investment and private ownership, neither of which was permitted elsewhere in the country. They charged low rates of taxation on profits made by foreign firms and reduced import duties. They concentrated on the manufacturing of low cost products for export and many foreign companies were attracted to these SEZs for routine manufacturing. Local leaders in the SEZs were permitted to follow economic policies which seemed at variance with Communist ideology in the interest of attracting further foreign investment. SEZs became the main driving force behind China's economic recovery.

Figure 100: The Special Economic Zones (SEZ) in China (The Gold Coast)

In 1989, in a further move towards encouraging economic cooperation with the West, Shanghai Stock Exchange reopened for the first time in 40 years and Shanghai is now a major global financial centre. The provision of banking facilities has further enhanced China's attractiveness to foreign investors. Deng Xiaoping died in 1997 but economic reform continued. Trade barriers were reduced, foreign and private investment were further encouraged. In 2001, China joined the World Trade Organisation (WTO), a move that helped establish China as a major world economy. Trade agreements were established with many western countries and regions, including the USA and the EU.

2. Large low wage labour supply

China's large population and low wages offered significant opportunities for large multinational companies looking to maximise their profits. These favourable conditions, along with the economic reforms, encouraged many multinationals to relocate some of their routine manufacturing to China. As the SEZs began to develop they attracted large numbers of rural migrants in search of paid employment. From 1990–2011, an estimated 250 million people moved from rural areas in the interior to the rapidly growing cities in the east. In 2013, rural migrants accounted for 40% of the urban workforce.

China's economic performance has been remarkable. It is largely based on the export of goods which have been assembled in China. The finished product still bears the brand name of the multinational company based in MEDCs. These include:

- electronic and telecommunications equipment, such as mobile phones and computers.
- clothes and shoes for many leading high street retailers in MEDCs.
- 'white electrical goods' such as washing machines and dishwashers.
- chemical fertilisers and other agricultural products.
- concrete, steel and other metals.
- computer games and toys.
- vehicle assembly, including luxury makes such as BMW and Mercedes-Benz.

Figure 101 lists some of the multinational companies operating in China.

Company	Country of origin	Product
Apple	USA	Consumer electronics – mobile phones, computers
Intel	USA	Consumer electronics – computers, silicon chips
Caterpillar	USA	Construction machinery and equipment
Boeing	USA	Aircraft manufacture
Proctor & Gamble	USA	Consumer goods – personal care, household cleaning products
Morgan Stanley	USA	Financial services
Goldman Sachs	USA	Investment banking
Yum! Brands	USA	Restaurant company – including KFC, Pizza Hut
Coca Cola	USA	Soft drinks
Seagate	USA	Technology – data storage, disk drives
Analog Devices	USA	Silicon chips
Monsanto	USA	Agrochemicals, agriculture biotechnology
DuPont	USA	Industrial biotechnology, chemical fertilisers, high yielding seeds and plastics
Whirlpool	USA	Home appliances, 'white electrical goods' – washing machines, freezers
Unilever	UK & Netherlands	Consumer goods, food and drink, personal care, household cleaning products
BP	UK and USA	Oil and gas energy supplies
Audi	Germany	Automobiles
Volkswagen	Germany	Automobiles
Samsung	South Korea	Consumer electronics – smart phones, home electronics, multi-media products
IKEA	Sweden	Self-assembly furniture

Figure 101: Selected Multinational Companies that operate in China

China's economic success

- China's GDP increased by an average of 10% annually from 1978–2013.
- China is the leading world producer of concrete, steel, ships and textiles.
- China produced close to 10 million cars in 2008 for the home market. This represented a 10 fold increase in numbers from 1992.
- A number of Japanese and other Asian car manufacturers have assembly plants in China. In addition, luxury brands such as BMW and Mercedes-Benz are also assembled for sale in China and South East Asia.
- Approximately 24% of the world's total exports of textiles are from China.
- China accounts for the bulk of the world's production in major consumer items, including 70% of the world's mobile phones, 60% of world's shoes, 50% of cameras and 25% of 'white electrical goods'.
- Three of the world's busiest ports are in China.

Figure 102: Numbers of middle class earners in major world regions 2013 (millions)

Source: Data from James Davies, Rodrigo Lluberas and Anthony Shorrocks, Credit Suisse Global Wealth Databook 2015

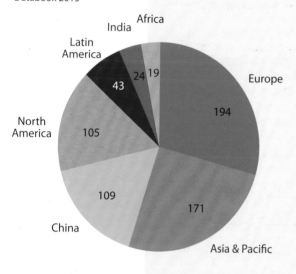

Recent developments

In the emerging markets the number of middle class earners has increased dramatically in recent years. (A middle class earner's income is set at twice the average national income.) In 2014, the number of middle class earners in China exceeded those in the USA (Figure 102). According to a recent report, over 90% of China's urban dwellers will be middle class earners by 2030. In addition, the proportion of affluent people is expected to increase (Figure 103). Crucially, this growing wealth will also create an internal demand for consumer goods. Already China has become the world's largest market for luxury cars and consumer goods, including designer fashion clothes, jewellery and cosmetics. This growing internal demand attracts more foreign investment.

Figure 103: Disposable Income* per capita in urban areas in China 2005–2015 (projected to 2030)

* Disposable income is income remaining after taxation

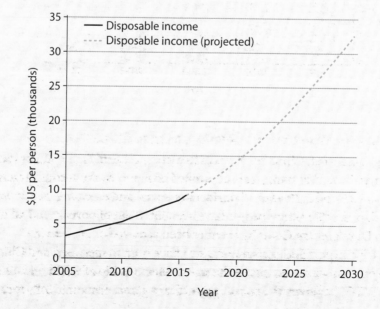

China's initial developments were associated with routine manufacturing jobs in branch factories, while the research and development of the products were carried out in the MEDCs. Increasingly, China is developing its own research and development capabilities, as well as moving into high-end consumer goods. The current Chinese President, Xi Jinping, is driving a 'Made in China' campaign to promote Chinese involvement in all stages of manufacturing. To achieve this, the government are investing in science and engineering research, and are offering favourable rates of taxation to firms to upgrade their production. Many Chinese students currently attend universities in the UK and USA where they will acquire the qualifications necessary to develop modern industry.

Exercise

1. Study Figure 104, which shows the proportion of China's urban households in selected income groups 2013–2015 and predicted values to 2030.

   (a) Describe the trends shown for each income group. Use figures in your answer.

   (b) How might these changes impact on future consumer demand in China?

Figure 104: Projection of China's urban households by income groups 2013–2015 (projected to 2030)

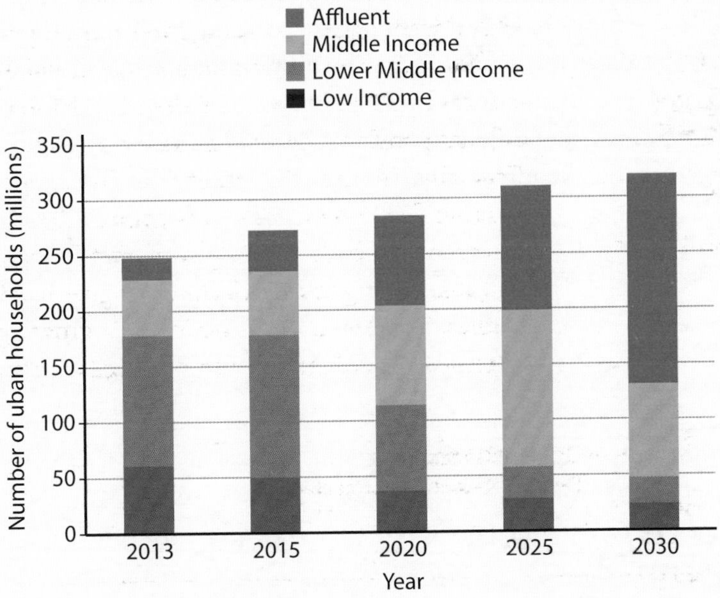

Issues arising from economic development in China

As an emerging market, China has been extremely successful. The country has moved from an inward looking, state controlled economy to the second largest world economy in under 30 years. Considerable social and economic development has occurred: over 500 million people have been lifted out of poverty and all of the Millennium Development Goals have either been achieved or are nearing achievement. Transport and communications have been modernised and China has some of the best railway networks in the world. Membership of world organisations, such as the WTO, has given the country significant global economic influence.

China has also achieved political acceptance by the West. The UK government's decision in 2016 to agree a deal that will use Chinese investment and technology to build a nuclear power plant at Hinckley Point is an example of this. By 2015, the rate of economic growth had slowed but China remains a major world economic power.

The economic benefits detailed in this study hide huge inequalities in wealth distribution. There are stark contrasts between the eastern urban areas and the interior rural areas, where many are still employed in traditional agriculture. Close to 100 million people, mostly rural dwellers are below the poverty line.

China has long been accused of serious human rights violations and many Western leaders have raised this issue with successive Chinese leaders at international meetings. During the 1980s, a movement seeking increased democracy was defeated, culminating in the well documented events in Tiananmen Square in 1989. Widespread censorship pervades all aspects of life in China. Internet activity is closely monitored and internet search engines such as Google have had to comply with Chinese censorship. International tourism has developed but tourism destinations and itineraries are strictly controlled by the Chinese Authorities.

Millions of people have been displaced in the interest of industrial development. Over 1.2 million were displaced to make way for The Three Gorges Dam on the Yangtze river. This dam is the world's largest hydro electricity supplier and will provide much needed power for China's growing industry. It will also improve navigation on the Yangtze river and Chongqing has become a major inland port on the Yangtze as a result of the development. In addition to the large-scale displacement of people, this project also raises major environmental and ecological issues. The dam is built in an area subject to earthquakes and dam building is thought to increase the likelihood of earthquake occurrence.

China's economic success has been based on the large-scale use of fossil fuels, which are a major source of the air pollution and greenhouse gases that contribute to global warming. Efforts have been made to improve this but air pollution remains a major issue in China. China has been expanding and investing in Africa, partly in search of raw materials, and many have voiced concern over this move (see pages 244–245). In addition, China has a low rate of population growth and in the near future will have to deal with the issue of ageing (see the impacts of the one-child policy on the population resource balance, page 169). In 2015, the one-child policy was abandoned but it remains to be seen how this will affect overall population growth.

Exam Questions

1. Explain what 'emerging markets' are in the context of LEDCs. [3]

   Question from CCEA AS2 Human Geography Specimen Paper, © CCEA 2016

2. With reference to your case study of an emerging market:

   - explain the factors that account for its economic growth.
   - describe the outcomes of this economic growth. [15]

References

China's Economic Miracle:

http://www.bbc.co.uk/news/world-asia-china-20069627

China's Special Economic Zones:

http://www.worldbank.org/content/dam/Worldbank/Event/Africa/Investing%20in%20Africa%20Forum/2015/investing-in-africa-forum-chinas-special-economic-zone.pdf

China's growing middle class:

http://uk.businessinsider.com/chinas-rising-middle-class-will-create-opportunities-the-world-has-never-seen-before-2015-5

http://www.telegraph.co.uk/finance/china-business/11929794/Chinas-middle-class-overtakes-US-as-largest-in-the-world.html

China's economic problems:

http://www.theguardian.com/world/2015/jan/25/china-bids-bring-economic-miracle-end-hard-times

http://www.theguardian.com/world/china

Glossary

Physical Geography

Abrasion: the scraping, scouring and wearing away of the bed and bank of rivers by rock fragments carried along by the river flow.

Afforestation: the deliberate planting of trees.

Alluvium: the term used to describe any sediment deposited by a river.

Annual hydrograph (regime): a graph which shows the variation in a river's discharge over one year. In Europe the water year normally runs from October to September.

Arcuate delta: a delta shaped like the letter 'delta' (Δ) in the Greek alphabet or a wedge of cake. The Nile delta in Egypt is one example.

Atmosphere: the layer of transparent gases that surrounds the earth held by gravity.

Attrition: the wearing down of the river load itself as particles strike each other and the bed and banks.

Base flow: the part of a river's discharge that is contributed by the groundwater store.

Billabong: an Australian name for an **oxbow lake** or meander cut-off.

Bird's foot delta: a delta in which the distributaries extend their channels and banks into the sea at the mouth of a river, such as the Mississippi in the Gulf of Mexico.

Braiding: the dividing of a river into several interweaving channels, normally the result of a loss of energy or overloading with sediment.

Channel catch: the term used to describe rain that falls directly into river channels.

Channelisation: the process of artificially modifying the natural channel of a river by changing its shape (re-sectioning), profile or course (realignment).

Confluence: the junction of two rivers.

Continentality: how the climate of a location is influenced by its remoteness from oceans and ocean air.

Convectional rainfall: precipitation resulting from air currents rising due to local warming. Short bursts of intense rainfall are a common outcome.

Corrasion: an alternative term for **abrasion.**

Corrosion: an alternative term for **solution** erosion by rivers.

Cyclonic rainfall: precipitation associated with low pressure systems and air rising at fronts where different air masses meet.

Deferred junction: where a tributary flows downstream alongside another river before its **confluence**. Also known as a **yazoo**.

Deforestation: the clearance of trees from an area.

Delta: the depositional landform built up from alluvium deposited at a river mouth.

Discharge: the volume of water passing a point in a given time. It is represented by the equation $Q = A \times V$, where Q is discharge, A is channel cross-section area and V is velocity. Discharge is measured in cumecs, cubic metres per second (m^3/sec).

Distributary: a river that branches from the main channel, commonly found in deltas.

Drainage basin: the land area drained by a river and its tributaries.

Drainage density: an index of the average length of river channel per unit area of a drainage basin (km/km^2).

Dredging: this is the underwater excavation of material from a channel bed to increase its capacity and its velocity by reducing friction.

El Niño: the name given to the periodic development of warm ocean surface waters along the Pacific coast of South America. The event is often linked to unusual and extreme weather conditions in other parts of the world.

Erosion: the wearing away of the land by natural processes, including ice, flowing water and wind.

Evaporation: the change of water from its liquid to its gaseous form. Along with transpiration it is an output from the drainage basin system termed evapotranspiration.

Flashy hydrograph: a water graph with steep limbs and a short time lag associated with the risk of river flooding.

Flocculation: the joining together of clay particles when fresh water meets salt water, for example, at a river's mouth.

Flood (storm) hydrograph: the term for a graph representing a river's response to a single rain event.

Gorge: a vertical sided river valley often formed by a retreating waterfall.

Hjulström curve: a graph that shows the velocity at which differing sizes of particle are eroded, transported or deposited by water flow (rivers).

Hydraulic action: a process of river bed and bank erosion; it involves the energy of the flowing water itself.

Hydrograph: a line graph or changes in river discharge over time.

Impermeable: a soil or rock type that does not let water pass into or through it.

Infiltration: the process of water soaking into the soil from the surface.

Interception: the process in which precipitation falls onto vegetation rather than directly to the ground.

Jet streams: these are narrow bands of fast flowing winds concentrated within the upper atmosphere. The polar jet stream in the mid-latitudes is one example.

La Niña: in effect the opposite of an El Niño. Under La Niña conditions, the Pacific trade winds become very strong and an abnormally large body of cold water forms in the central and eastern Pacific Ocean basin.

Levées: raised river banks. They may be formed by deposition of larger alluvial particles during floods but are often artificially raised to prevent such floods.

Load: the term used to describe any sediment carried by a river.

Meander: a bend in a river.

Mort lake: another term for a **oxbow lake** or cut-off meander.

Orographic or relief rainfall: precipitation caused by air rising over upland areas causing cooling and condensation.

Overland flow: water run-off over the land surface but not in a channel.

Oxbow lake: the remnant section of unused channel when a river changes its course, cutting off a meander.

Peak discharge: the period of highest river water flow during a storm or flood episode.

Peak rainfall: the period of most intense precipitation during a storm.

Percolation: the downward movement of water from the soil into the deeper stores such as ground water.

Permeable: a soil or rock type that allows water to pass into and through it.

Plunge pool: the deep section of river bed immediately below a waterfall.

Point bar deposits: the alluvial sediments left on the inner bank of a meander by slow moving water. They form the slip-off slope.

Precipitation: the name for any form of water leaving the atmosphere, including rain, hail, sleet, snow, dew, fog and frost.

Rainfall floods: inundations of land due to direct heavy rainfall exceeding the infiltration capacity and not rivers simply overflowing.

Recession limb: the declining slope following peak discharge on a hydrograph.

Riffles and pools: these are the regular pattern of shallow and deep sections of channel that form in river beds.

Rising limb: the increasing curve on a hydrograph leading to peak discharge.

River floods: inundation of land as the result of a river channel overflowing.

Saltation: the process by which material is transported in a bouncing motion, such as bed load moving downstream in a river or sand across a beach.

Snowmelt: the release of water when winter snows melt in spring, often causing flooding.

Solution: the term used for both an erosion process and the transportation of material chemically dissolved in water.

Stores: in a system these are the places to and from which material and energy is transferred.

Storm flow: describes all the river discharge resulting from a period of rain. It is graphically represented as the area under the curve in a hydrograph above base flow.

Suspension: the river transport process where light particles are held and carried in the flowing water.

Synoptic chart: a map summarising the atmospheric (weather) conditions across an area.

Systems theory: a framework used to describe and analyse the interaction of parts of the physical and human world. Examples include ecosystems, drainage basins, factory production and farms.

Thalweg: the line of maximum velocity in a river channel.

Traction: a river transport process involving bed load material rolling downstream.

Transfers: exchanges of material and/or energy between stores in a system.

Transpiration: the loss of moisture from vegetation during photosynthesis.

Transport: in fluvial studies, this is the movement of material downstream.

Tributary: a river or stream that joins and adds (*contributes*) water to another channel.

Throughflow: the term used to describe water moving downhill within the soil.

Tundra: a high latitude biome with a limited productivity and range of species due to its extreme climate.

Upper Westerlies: persistent upper troposphere winds that flow west to east around the mid and high latitudes.

Water table: the height of water stored as groundwater within rocks beneath the surface of a drainage basin.

Waterfall: a steep or vertical section of channel where the river falls freely.

Watershed: the line that marks the boundary between drainage basins.

Yazoo: a term used for a **deferred junction** on a river. The name comes from an example on the Mississippi floodplain.

Human Geography

Adult literacy rate: the percentage of adults who can read and write.

Agenda for Sustainable Development: provides an action plan that will shape global development between 1 January 2016 and 31 December 2030. It follows the Millennium Development Goals (MDGs) programme.

Agribusiness: the industrialisation of agriculture is an example of globalisation.

Aid: the transfer of resources from rich countries, usually MEDCs, to poor countries, usually LEDCs.

Areas of Outstanding Natural Beauty (AONBs): these are usually smaller in area than National Parks. There are 46 in all, covering about 15% of England. Their management is the responsibility of the local council.

Bilateral aid: aid that comes directly from one country to another.

BRICS: this is an acronym for Brazil, Russia, India, China and South Africa. BRICS are the longest established and most successful of the emerging markets. A BRICS economy is defined as an emerging market having 3% or more of global GDP.

Brownfield sites: derelict sites, usually in inner cities, which can be redeveloped.

Colonialism: where one country takes political control over another, usually as part of empire building.

Composite measures of development: those measures that use several indicators to provide an index of development which can then be used to rank countries.

Counterurbanisation: a movement of urban workers from a city to rural towns and villages within commuting distance of the city.

Crude Birth Rate* (CBR): the number of live births in a year per 1,000 of the mid-year population.

Crude Death Rate* (CDR): the number of deaths in a year per 1,000 of the mid-year population.

> ** If the difference between these two measures is positive there is an increase in population (**natural increase**) and conversely if the difference is negative, there is a population decrease (**natural decrease**).*

Debt: the amount of money owed, usually by a LEDC, to another country or organisation.

Demographic Transition Model (DTM): a model that charts changes in birth and death rates and population growth over time.

Dependency ratios: the ratio of economically active population to the economically inactive or dependent population.

Aged Dependency is calculated:

$$\frac{\text{Total number over 65}}{\text{Total number 15–64}} \times 100$$

Youth Dependency is calculated:

$$\frac{\text{Total number 0–14}}{\text{Total number 15–64}} \times 100$$

Emerging Markets: a group of former LEDCs that have made rapid economic progress. They are associated with low production costs and produce goods for the export market. They were first referred to by an economist at the World Bank in 1981 but have become popularised by the investment bank Goldman Sachs since 2001.

Enrolment in primary, secondary and tertiary level education: the percentage enrolment in each stage of education.

Epidemiological transition: shows how causes of death change over time and at different levels of development.

Fertility policies: government policies to control birth rates. Anti-natalist policies aim to reduce the number of births. Pro-natalist policies aim to increase the number of births.

G20: a forum of the world's major economies and banks. G20 includes USA, Russia, China, UK, EU, India and Mexico.

Gentrification: an outcome of urban regeneration in inner cities where rebuilding and restructuring result in the area becoming more affluent. Often the former inhabitants are unable to afford the new housing and lack the skills needed for the new jobs.

Global Goals for Sustainable Development: 17 goals which will deliver Agenda 2030. These goals will be monitored across 169 targets.

Globalisation: the current interaction of most of the world's economies. In the world of today national economies are no longer separate entities but rather part of a world or global economy.

Greenfield sites: rural land that is being developed for non-rural use.

Gross Domestic Product GDP per capita (GDPpc): an economic measure of development. It is the total value of goods and services produced in a country. In order to allow international comparisons to be made it is always expressed in US dollars.

Gross National Income per capita (GNIpc): formerly known as GNPpc, GNIpc is the total value of goods and services produced in a country, plus taxes and income from abroad in one year, divided by the total population. In order to allow international comparisons to be made it is always expressed in US dollars.

Heavily Indebted Poor Countries (HIPC) Initiative: is a debt reduction scheme set up in 1996 for heavily indebted poor countries that satisfy economic and political criteria laid down by the IMF and World Bank.

Human Development Index (HDI): measures the average figures for life expectancy; mean years of schooling for those aged 25 and over and expected years of schooling for those of school entering age; and Purchasing Power Parity (PPP). The values of HDI fall in the range 0–1 and countries are ranked according to their calculated value.

Infant Mortality Rate: the number of live children per 1,000 who die within the first year of life.

Informal settlements: housing built without planning permission on any available piece of land in LEDCs, usually without any basic infrastructure. These settlements are built using whatever materials are available – corrugated iron, timber, even plastic sheeting. The people living in these informal settlements have no legal right to occupy the land and local authorities do forcibly remove them on occasions.

Life Expectancy at birth: the number of years a person is expected to live calculated at the time of their birth.

Maternal Mortality Rates (MMR): the number of maternal deaths per 100,000 live births.

Millennium Development Goals (MDGs): these consisted of 8 goals devised by the UN to steer development in LEDCs from 2000–2015. Each goal sub-divided into a number of targets. Progress on each target was monitored by a series of indicators.

MINT: an acronym for Mexico, Indonesia, Nigeria and Turkey. The MINT grouping was formed in 2013 and includes those emerging markets countries whose economies have between 1.3–2.9% of global GDP.

Multilateral aid: aid that comes from several countries or organisations such as the World Bank.

Multilateral Debt Relief Initiative (MDRI): a scheme devised by the World Bank and other international organisations in 2005 to cancel all debt in some of the poorest countries to help them achieve the MDGs target of poverty reduction. Tanzania benefitted from this scheme in 2006.

Multinational Company: divides the running of the company across several countries.

National Census: a count of all of the population and those social and economic characteristics that can easily be counted on a specific date and usually every ten years.

Natural population change: the balance between birth rates and death rates.

National Parks: a form of protected land. The first National Parks were set up during the 1950s to manage conservation and recreation. There are currently 15 National Parks in the UK. Each Park has a National Park Authority whose responsibilities include conservation of the natural beauty, wildlife and cultural heritage of the Park while at the same time improving opportunities for public understanding and enjoyment of the Park.

Official aid: aid that comes directly from government sources.

Optimum population: the maximum number of people that can be supported in a region without causing environmental damage or over using resources for future generations. **Overpopulation** refers to situations where there are too many people for the available resources. **Underpopulation** means there are too few people to utilise the available resources.

Population sustainability: a situation where the population is not over using resources or causing adverse environmental effects.

Purchasing Power Parity (PPP): this takes account of the real purchasing power of a given amount of money in different countries and as such is a more reliable measure than GNIpc for LEDCs.

Re-urbanisation: regeneration or revival of declining inner urban areas closely associated with gentrification.

Rural-urban continuum: the gradual change from urban land use to rural land use.

Rural-urban fringe: the zone where urban and rural land use meet.

Sites of Special Scientific Interest (SSSIs and ASSIs in Northern Ireland): these are areas that have special wildlife or rare flora. There is a list of restrictions in force in SSSIs and applications for development within a SSSI must be passed by the SSSI regulators. Most are in private ownership.

Suburbanisation: a process that refers to the decentralisation of people, services and industry to the edge of the existing urban area (urban sprawl).

Total fertility rate (TFR): the average number of children a woman will have during her reproductive years (15–44 years) per year assuming she will live to the end of her reproductive life. A TFR of 2 is referred to as **Replacement Level Fertility**.

Trade: the exchange of goods between one country and another.

Transnational Company: a company that has its headquarters in a single MEDC and a number of branch factories in LEDCs.

Vital Registration: the official recording of all births, including stillbirths, adoptions, marriage and civil partnerships, and deaths. In Scotland there is also information on divorce.

Voluntary Aid: aid that comes from charitable organisations such as Oxfam, Trócaire and Action Aid. These are referred to as **Non-Governmental Organisations (NGOs)** and they rely entirely on voluntary contributions.

Copyright

Acknowledgements

Questions from CCEA Geography Past Papers, 2010–16 are included with the permission of the Northern Ireland Council for the Curriculum, Examinations and Assessment, © CCEA 2016.

Credits

The following photographs, diagrams, maps, graphs and tables are all included with the kind permission of the copyright holders. The numbers denote page numbers.